# Test and Evaluation of Complex Systems

# WILEY SERIES IN MEASUREMENT SCIENCE AND TECHNOLOGY

**Chief Editor**
**Peter H. Sydenham**
*Australian Centre for Test & Evaluation*
*University of South Australia*

---

# Test and Evaluation of Complex Systems

Matthew T. Reynolds

JOHN WILEY & SONS

Chichester · New York · Brisbane · Toronto · Singapore

Copyright © 1996 by John Wiley & Sons Ltd,
Baffins Lane, Chichester,
West Sussex PO19 1UD, England

*National*     01243 779777
*International*  (+44) 1243 779777
e-mail (for orders and customer service enquiries): cs-books@wiley.co.uk

Visit our Home Page on http://www.wiley.co.uk
or
http://www.wiley.com

*Other Wiley Editorial Offices*

John Wiley & Sons, Inc., 605 Third Avenue,
New York, NY 10158-0012, USA

Jacaranda Wiley Ltd, 33 Park Road, Milton,
Queensland 4064, Australia

John Wiley & Sons (Canada) Ltd, 22 Worcester Road,
Rexdale, Ontario M9W 1L1, Canada

John Wiley & Sons (Asia) Pte Ltd, 2 Clementi Loop 02-01,
Jin Xing Distripark, Singapore 0512

*Library of Congress Cataloguing-in-Publication Data*

Reynolds, Matthew T.
    Test and evaluation of complex systems / by Matthew T. Reynolds.
        p.   cm. — (Wiley series in measurement science and technology)
    Includes bibliographical references and index.
    ISBN 0 471 96719 X
    1. Systems engineering.   2. Testing.   3. Evaluation.   I. Title.   II. Series.
TA168.R49   1966
620'.0044 — dc20#96-28134
#CIP

*British Library Cataloguing in Publication Data*

A catalogue record for this book is available from the British Library
ISBN 0 471 96719 X

Typeset in 10.5/12.5pt Times by Tradespools Ltd, Frome, Somerset
Printed and bound in Great Britain by Bookcraft (Bath) Ltd
This book is printed on acid-free paper responsibly manufactured from sustainable forestation,
for which at least two trees are planted for each one used for paper production.

to Brenda, Theresa,
and Patrick –
my inspiration for everything

## Cover Illustration

This family of instrumented dummies, representing the range of human sizes and conditions, is typical of those used in the crash testing of new automobile designs. Photo compliments of First Technology Safety Systems of Plymouth, Michigan, U.S.A.

# Contents

# Preface

Test and Evaluation has come of age. In the development of complex systems and products, T & E has now taken its place not only as an integral part of every phase of systems engineering, but also as a major metric of program management, helping to identify and assess through measurement the cost, schedule, and performance risk at every level of activity. It can no longer be pictured as the last block in a design-build-test series of activities leading to putting a new capability into service.

The second half of the 1800s saw an explosion of invention – everything from the typewriter to the automobile – that permanently transformed mankind. We are now in another age of great progress. It has been said that in the 1960s, technology crossed a threshold in applying technology. We started building things we could not fully test. In the 1980s, it crossed another threshold: we could not build things without some significant human error. Without a change in the role of T & E, we could not be assured of being able to build systems that even worked properly. With good, affordable modeling and simulation available to supplement T & E, we are now able to offset the limitations due to human error. And with so many computer aided design and T & E tools, our confidence in the validity of the T & E processes and procedures themselves can be measured and maximized. The new age of technological progress is fostering a new – perhaps the first – age of T & E.

This book provides a framework within which to understand this new role of T & E and gives some recent examples of how T & E has successfully supported today's complex development programs. It has been written first as a textbook to help promote the study of T & E, and secondly as a compendium of 'lessons learned', since that is largely how T & E has evolved.

# Series Editor's Preface

The Series provides authoritative books, written by internationally acclaimed experts, on the topics that support measurement science and engineering.

Test and Evaluation (T & E) is the name given to the measurement process that is used to assess, with known confidence, the operational effectiveness, suitability and performance of complex engineering systems – from conception through to disposal.

This book provides a universal framework within which the application of the new T & E wisdom can be applied – to mitigate program risks, to measure program success and to improve confidence that the right decisions are being made throughout the development and through-life modification of complex engineering systems.

This title, the work of a pre-eminent T & E practitioner and lecturer, is the first to model the rapidly evolving T & E process in a universal framework and serves as both a text for the novice and a desk-top guide for the expert.

As a world first in this series, it is the cornerstone upon which further research and investments can be built – by academia, by industry and by Government – locally, nationally and internationally.

**Peter Sydenham**
Editor-in-Chief

# Foreword

The practice of testing and evaluation has been an integral part of human activity from earliest times. The Old Testament describes the sequence of tests conducted by God to support His evaluation of Job's spirit. Even earlier was the ill-fated test flight of his Dad's (actually Daed's) synthetic wings conducted by Icarus, whose trajectory veered out of the envelope and too near the sun; that test event should clearly have been rescheduled to the night shift. Of perhaps greater notoriety in history, however, have been the tests not performed and the evaluations never completed. One thinks of the customer acceptance testing omitted on the Trojan Horse and the pre-construction soil tests for the Upright Tower of Pisa, cancelled probably due to scheduling and budgetary pressures.

Much of the impetus for the development of the test and evaluation discipline as it is practiced today has, in fact, sprung from its crucial role in the creation and procurement of complex and costly military, space and transportation systems. In current times, it would be unthinkable for an organization to dedicate significant resources to the construction of a new airplane, automobile or submarine without also planning and conducting an adequate program of testing and evaluation. Such testing would address hardware, software and people, and would employ techniques and instrumentation rivaling in sophistication those of the object under test. For example, advances in computer science now permit *virtual* test and evaluation, or what I like to call 'apparitional testing', defined as 'testing performed on non-existent objects and processes, by people who are not present, using resources which cannot be seen'. In any event, testing is usually conducted for one or more of the following reasons:

- To find out interesting things about the subject system [Experiment],
- To prove to somebody that the system works [Demonstration], or, perhaps most importantly,
- To reduce the risk that the system will fail when placed in service [Job Security].

In addition to minimizing performance risks related to effectiveness and suitability, appropriate test and evaluation can also help reduce programmatic risks involving cost and scheduling, as well as risks due to unintended events, e.g., safety concerns.

As technology has advanced, and systems exploiting it have become more complex, test and evaluation has evolved from an art to a craft to a scientific and engineering discipline. Unfortunately, the formal literature has not kept pace; there are very few books which address the process and substance of test and evaluation to any degree of generality. Thus Matt Reynolds' treatise is a unique contribution to the field and a real gift to both student and practitioner. Matt's professional credentials are impeccable, embracing decades of teaching, writing, and 'doing'. Anyone with an interest in the test and evaluation of complex systems will enjoy this book and benefit from reading it.

**Donald R. Greenlee**
President
International Test and Evaluation Association
Fairfax, Virginia
January 25, 1996

# Acknowledgments

The author would like to acknowledge the helpful assistance of the following organizations in the preparation of this book. They provided photos and information about their programs, in the interests of sharing the lessons they had learned in the course of their development programs. The Federal Aviation Administration also provided the author an excellent tour of its facilities at the FAA Technical Center in Atlantic City, New Jersey, and the Insurance Institute for Highway Safety provided one of its impressive facilities at the Vehicle Research Center in Ruckersville, Virginia.

Boeing Commercial Airplane Group, Seattle, Washington
Calspan Advanced Research Center. Buffalo, New York
Chrysler Corporation, Highland Park, Michigan
First Technology Safety Systems, Plymouth, Michigan
Fisher Price Toys, Aurora, New York
Forbairt Technology Services, Dublin, Ireland
Ford Motor Company, Global Test Operations, Dearborne, Michigan
General Electric Aircraft Engines, Product Test Center, Cincinatti, Ohio
Insurance Institute for Highway Safety, Arlington, Virginia
Nevada Automotive Test Center, Carson City, Nevada
Underwriters Laboratories, Northbrook, Illinois
US Federal Aviation Administration Technical Center, Atlantic City, New Jersey
US Department of the Interior, Minerals Management Service, Herndon, Virginia
US Department of the Navy, Naval Undersea Warfare Center, Keyport, Washington
Wyle Laboratories, Norco, California

# 1

# The New Age of Testing

*One test is worth a thousand expert opinions*

Everyone is interested in testing. Perhaps it is more accurate to say that everyone is interested that proper testing is conducted and that the test results are objectively evaluated. We may not think of it often, but we assume that everything receives an adequate amount of test and evaluation (T & E). We expect that our automobiles have been tested and evaluated to assure good performance, that machines that process our food have been tested and inspected to be clean, and that the airplanes we fly have been tested and evaluated to be safe. We not only expect that such T & E is done, but we are willing to pay the costs associated with it.

Simple testing is easily understood. When we use the term 'testing', we intuitively understand that it includes the planning of the test, the conduct of the test, and the evaluation of the results. In complex systems, however, the evaluation of results has in itself become so specialized that by common practice, it has now been given its own visibility. The 'evaluation' part of Test and Evaluation has come to encompass the acquisition of adequate instrumentation and other test facilities, the collection of data, the analyses of the data, and the assessment of the results. In this book, we will use the term Test and Evaluation (T & E) in referring to all of these efforts.

Testing is a very common activity. We test the feel of fruit in a produce market to help determine its freshness. We test an electric lamp after we have rewired it to make sure it works. We take an automobile for a test run before considering buying it. But technology is transforming our world at ever increasing rates. Lamps and automobiles are not as simple as they were forty – or even ten – years ago. Testing and evaluating them is not as simple either. Aside from an incredible array of fancy consumer products, technology has also provided us with an ever-increasing multitude of systems that provide services to groups of people, to cities, to national governments, and indeed to the entire world. Through the pursuit of better lifestyles, greater freedom, and more leisure time, these systems and products have greatly complicated our lives. They have made us more dependent on each other and on our governments. They have also made us more sensitive to the disastrous consequences that can accrue when these systems do not work as advertised, or fail us at the worst possible time.

Testing has always been done. Adam and Eve surely 'tested' the fruit of the forbidden tree to make sure they had fresh pieces. But T & E has come to be recognized as a scientific discipline only within the last few decades. The drive to identify a set of common principles for T & E and for the successful approaches to the management of T & E programs has taken hold only since huge airplanes have crashed during take-off, automobile designs have been found to have serious defects, tanks have not worked on an exercise battlefield, the telephone system network covering the entire east coast of the United States went down, electricity was lost in several states for almost 24 hours, and a space shuttle exploded during lift-off. This book describes the basic principles that have come to underlie the planning, execution, and reporting of Test and Evaluation programs.

## 1.1  TEST AND EVALUATION DEFINED

T & E is the measurement of the performance of a system, and the assessment of the results for one or more purposes. T & E is conducted to help make engineering programmatic or process decisions, and to reduce the risks associated with the possible outcomes of those decisions. The specific purposes of a particular T & E program are an important driver of its character and scope. Figure 1.1 is a list of different types of purposes for which T & E is conducted. A typical T & E program or a particular test event within a program would involve a mix of several of them. It would appear that T & E defies boundaries and categorization – that it is an art more than a science. In fact, much has been done in the last two decades to characterize what T & E is, develop guidelines for planning and conducting it, and capture the lessons learned that experience tells us are critical to having successful T & E programs. This book describes the principles that have emerged.

## 1.2  T & E IN THE SYSTEMS ENGINEERING PROCESS

The Industrial Revolution brought about what has been called the Machine Age. Machines were designed to do work previously done by people. The 1940s brought about what can be referred to as the Systems Age. Large-scale systems have become more complex and more common: they typically include an interrelated combination of materials, equipment, computer software, facilities, data, money, etc. designed to perform some significant set of operations. General

- Prove a concept

- Ensure safety

- Ensure adequate human factors engineering
  (man/machine interface)

- Ensure user requirements are met

- Avoid failures in service

- Ensure contract compliance

- Ensure that fixes & enhancements work

- Support acquisition (investment) decisions

- Provide feedback to the designer

- Verify supportability

- Validate models & simulations

- Compare systems

*to reduce risks*

**Figure 1.1**
Typical objectives of T & E

systems theory has provided a framework for organizing new science and engineering disciplines, which were and are ever-increasing in number because of rapidly advancing technologies. Applying systems theory to new system or product developments recognizes that the whole can have different operating characteristics than simply the sum of the parts, and that the difference between the whole and the sum of the parts can be advantageous (it can *add* effectiveness and efficiency to the aggregate effectiveness and efficiency of the various parts) or it can be disadvantageous (it can *detract* from the aggregate). Hence, it is not only worthwhile but also necessary to consciously design at the whole system level in an iterative process that interacts with the design of the individual parts, the subsystems level. This process is called systems engineering.

Broadly defined, systems engineering is the effective application of scientific and engineering efforts to transform a desired capability into a defined system design. The systems engineering process is a top-down iterative effort involving requirements definition, functional analysis, synthesis, optimization, design, and T & E.

What distinguishes systems engineering from earlier engineering development approaches is its emphasis on: (1) the use of a top-down approach in conjunction with the bottom-up design approach historically used; (2) life-cycle considerations, such as operation, maintenance, spare parts support, and disposal, in addition to development and production; (3) a more rigorous effort to identify system requirements and to refine them throughout the program; and (4) an integrated approach to ensuring an appropriate balance of the ever-increasing number of interdisciplinary engineering efforts and smart trade-off decisions made during the design (Blanchard 1991).

   T & E today plays an important part in every phase of the systems engineering process:

- confirming the need: marketing T & E is sometimes conducted on a system or product already in use, but reconfigured to represent selected features of a system upgrade under consideration.

- setting requirements: developmental T & E, in combination with modeling and simulation, is used to select the T & E strategy, to scope the program to determine what is affordable, and to identify which are the lower risk alternative approaches to achieving the objectives.

- design of the subsystems and interfaces: T & E is conducted integral to the systems engineering of the subsystem designs and of the interfaces. It is becoming more common not only that systems are subjected to developmental T & E to verify that they meet their performance requirements in the specifications, but also that they are subjected to operational T & E to verify that they will work in the user's environment, that they are interoperable with other systems, and that the customer will be satisfied.
- production: T & E is used to confirm adherence to specifications and verify adequate quality.

   A simplistic view of how T & E fit into the systems development process several decades ago could be depicted as in Figure 1.2. On the other hand, a more updated version of the role of T & E in today's development programs is depicted in Figure 1.3.

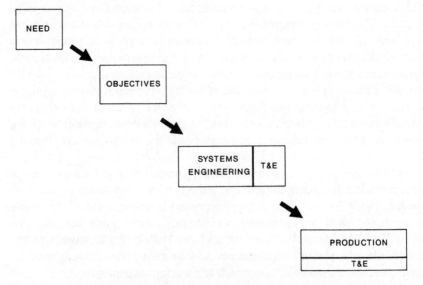

**Figure 1.2**
Old model of T & E in development programs

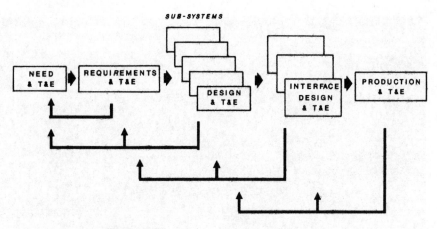

**Figure 1.3**
T & E in the system development process today

## 1.3   DRIVERS OF T & E

People want credible T & E programs. Years ago, there was little awareness of or interest in T & E. However, consumers and the public in general have been sensitized to what happens when T & E does not identify serious problems before a product or system is placed in service. And within the last few decades, government and industry organizations have started publishing many design regulations and standards, the enforcement of which requires T & E.

Pressures from consumer groups for better safety in areas such as automobile design and children's toys have focused much interest in the last two decades on how industry tests its products. It is not unusual to see impressive, graphic crash tests as part of the marketing material used by new car manufacturers. Besides consumer groups, insurance companies also are pressuring for better testing. At least one consortium of insurance companies is actually conducting its own testing. The Insurance Institute for Highway Safety, a coalition of property casualty insurance companies, has built a Vehicle Research Center in Ruckersville, Virginia. This 8.5 million dollar facility, opened in December 1992, is used for the study of crash test results. Most models of cars are crash tested at the facility, and the results are made available to consumers, to insurance companies, to state and national governments, and to the media. Engineers from the Center investigate actual car crashes to correlate the results with those from their tests. The results of research at the Center are also of interest to the auto manufacturers who compare them with their own crash tests. Figure 1.4 shows the crash hall, to which are attached two 600 foot runways and a 150 foot runway. Figure 1.5 shows the author inspecting damage on one of the vehicles purposely crashed.

**Figure 1.4**
Crash hall at the Vehicle Research Center

In children's toys, Fisher-Price has become known as an industry leader, not only for safety but for overall quality. Fisher-Price conduct extensive safety tests, use and abuse tests, and tests to determine the 'play value' of each new toy before they produce it. It is well known by anyone who has had children in the last several decades that Fisher-Price toys endure. They survive not only the original users, but are usually passed on and on to children of later generations.

Another important factor driving the interest in T & E today is the risk of consumer litigation. *MacPherson* vs. *the Buick Motor Company* in 1916 was a landmark legal case in determining liability suits. The plaintiff had been injured when a wheel collapsed on the automobile he was driving. Buick based its defense on the fact that it had sold the car to a dealer and had no contact with the injured party. The dealership claimed it was not liable because it did not manufacture the car. The judge ruled that Buick was indeed responsible. He stated that the manufacturer had a duty to inspect its products for defects, and that failure to do so constituted negligence (Hammer 1972). Over the years, the courts have greatly reduced the need for an injured person to show negligence on the part of a manufacturer, assembler, or retailer. The burden of proof has shifted from the user to the designer (Bailey 1989).

Underwriters Laboratories, Inc. (UL) conducts a very successful business that owes its success to consumer concerns about safety (Barrett 1992). UL charges

**Figure 1.5**
Author inspecting crash-tested automobile

manufacturers a fee to test their products to the safety standards that UL develops. The photo in Figure 1.6 shows UL testing of a life preserver and that in Figure 1.7 is of a gas fireplace insert. Manufacturers pay the fees because the UL tag is a priceless marketing tool. Now over 100 years old, UL was founded by an electrical inspector who set up a small testing lab a few miles from Chicago's 1893 World's Columbian Exposition. The expo exhibited over 100,000 Edison incandescent light bulbs and highlighted the beginnings of electricity, for which safety was considered a concern. Today, UL has almost 200 inspection centers in 63 countries. It has expanded its operations from testing things like toasters, television sets, and wiring to make sure they don't catch fire, blow up or give off shock to testing toilet paper brands to see if they are biodegradable and otherwise environmentally sound.

Oversight of commercial products by US government regulatory agencies has also increased attention to T & E. For example, the US Food and Drug Administration must approve the T & E results for new drugs before they are marketed to ensure that they are effective and safe, and must even approve that the planned T & E with actual people is safe before it proceeds. Over the past two

**Figure 1.6**
UL testing of a life preserver

decades, the Surgeon General of the United States has forced much T & E of the harmful effects of cigarette smoking. Testing to determine the effects of aerosol sprays on the ozone layer has also been underway for almost two decades.

In the US government's own programs, safety has also had a major influence on T & E. In the investigation into the accident of the Space Shuttle CHALLENGER, the public was given great insight into the extensive safety precautions that the National Aeronautics and Space Administration (NASA) uses in its engineering and test programs. In Defense Department programs, Congress has enacted specific legislation to require realistic 'live-fire' testing of munitions against manned vehicles (tanks, aircraft, ships) to assess the vulnerability of the crews to injury.

Safety has been only one driver in US government T & E. Providing as much if not more pressure than safety considerations has been the financial risk that a system may not meet all of its performance requirements when finally produced and placed in service. The systems that the federal government buys today are very complex, and represent a large investment of public funds. It was a major embarrassment to NASA when defects in the mirrors of the Hubble Telescope were not uncovered by the T & E program conducted prior to launching it into space in 1992. Two years later, an expensive Space Shuttle mission had to be launched to fix the problems. With today's computer driven systems, many with hundreds of millions of lines of computer code, it is possible that some significant design errors can go undetected, even over the course of many years of actual system operation.

**Figure 1.7**
Testing of a gas fireplace at UL

Among government programs, it is the military programs that are the most complex, and that present the most difficult challenges in planning T & E, for three reasons. First, specific requirements for the typical system are constantly changing. A system under development is being designed to protect against both threats that are current but about which knowledge is incomplete and threats that are postulated well into the future. Those threats – and the assessments of those threats – are ever-changing. Secondly, military programs are very technically complex, trying to take advantage of the most up-to-date technologies to gain critical nanosecond advantages in reaction times and accuracies that could easily make the difference in the outcome of a battle engagement. Thirdly, the user of the system (soldier, sailor or airman) is far removed from the people who define and articulate the requirements and the schedule, and who allocate the resources available for the program. Generally, those requirement setters are in Washington – removed geographically, and to an extent culturally, from the user. What Wilton P. Chase said in *Management of Systems Engineering* in 1974 is particularly true of development programs today (Chase 1974):

> The transition from understanding the approach to hardware-oriented designs to that for understanding that for system-oriented ones requires the recognition that all equipment intended for system application is being designed to be used and supported by a customer with a set of needs and values by the satisfaction of which he will judge the utility of the end product. Consequently, the system designer must be responsible for the integrated design of all elements of a system in relation to meeting the customer's specified operational use requirements, whether they be explicit or implicit.

For many military systems, there are substantial risks in development; risks in production; risks in their capabilities being optimized on the battlefield by soldiers, sailors and airmen with an average ninth grade education; and frequently even risks of safe disposal at the end of their useful lives.

**Figure 1.8**
Updates of the U.S. Air Traffic Control Equipment undergoing T & E at the FAA Technical Center

*Case example: aviation and avionics systems*

One example of an area where there is significant T & E in both government and industry – for many of the reasons discussed above – is that of aviation and avionics systems. The US Federal Aviation Administration (FAA) conducts significant government T & E of the systems it develops and operates itself, such as Air Traffic Control systems, as well as what industry develops and operates, such as aircraft and avionics. In addition, it publishes standards for companies to use in their T & E.

The Federal Aviation Administration Technical Center in Atlantic City, New Jersey, conducts T & E in the areas of air traffic control, communications, navigation, airport and aircraft safety and security. The 5000 acre facility includes a full-scale fire-test facility, wind tunnels, a chemistry laboratory, engine test cells, fuel safety test facilities, and a drop test tower to measure vertical impact. In addition, T & E is conducted in an air traffic simulation facility and a $150 million air traffic control system support computer complex. Future automated air traffic control systems are developed by the FAA and undergo T & E at the Technical Center (see Figure 1.8) before implementation in the field. The Center maintains, modifies and operates a fleet of specially instrumented test-bed aircraft to support airborne development and T & E programs. These 'flying labs' are used to test new airborne equipment and operational procedures, and to flight-check experimental ground-based navigational aids and guidance systems. Figure 1.9 shows testing at the Indoor Fire Testing Facility to determine the effectiveness of fire-blocked aircraft seat cushions and materials. Designed to withstand a 20 square foot jet-fuel fire, the building houses a main test bay, a computer control room, electrical/mechanical shop and a storage facility. Instrumentation measures temperature, heat flux, smoke density, and certain gas concentrations, such as carbon monoxide, carbon dioxide, and oxygen. Post-crash and in-flight scenarios conducted in the facility have been critical in developing a number of fire safety regulations. Wind tunnels, a pressure chamber and an altitude chamber are used to simulate flight conditions for components used in aircraft fire tests.

Two additional and related reasons why even some governments have made investments in T & E is to encourage technology development and to increase the markets for new products. Since 1978, the Government of Ireland has operated a National Electronics Test Centre (NETC) in Dublin to conduct electrical, electronics and communications systems T & E as well as to provide consultant services to companies conducting their own T & E. NETC is now part of Forbairt, the government agency established in 1994 to facilitate the development of Irish industry and to encourage technology and innovation.

**Figure 1.9**
Indoor Fire Testing Facility at the FAA Tech Center

The Centre conducts T & E to verify conformance to national and international standards (see section 4.3.1 for more on T & E standards), to assist in product development, and to uncover problems related to safety, environmental hardness (such as temperature, humidity, vibration and mechanical shock), and electro-magnetic compatibility. Figure 1.10 shows testing at the NETC lab. The need for and value of organizations such as NETC are growing rapidly because of the expansion of many business products to the 'global marketplace.' For this purpose, the staff of the NETC participates in many European Union T & E standards projects such as those of the European Commission Conformance Testing Services, and in working groups of the European Telecommunications Standards Institute. NETC's laboratory is accredited for its T & E by European organizations such as the European Commission, allowing it to qualify Irish products for sale in Europe as well as to qualify other countries' products for sale in Ireland. Its customers include companies in the United States, the United Kingdom, Germany and Japan. Through accreditation, products tested by NETC can be qualified for approvals such as the European Union's CE Mark, the VDE in Germany, and the UL tag in the US.

**Figure 1.10**
Testing at NETC's laboratories in Dublin, Ireland

## 1.4  SUMMARY

The rate of technology advances has forced much more attention on T & E than it had in the past. Consumers insist on good T & E for the products they buy. In some areas, governments and industries have set T & E standards for systems and products. The US federal government has set very strong T & E policies for many of the systems and products it develops for functions it performs, such as military systems and national air traffic control equipment.

## REFERENCES

Bailey, R. W. (1989) *Human Performance Engineering*. 2nd edn. Prentice-Hall, Englewood Cliffs, N.J.

Barrett, W. F. (1992) Testing for Money. *Forbes Magazine*, July 6.

Blanchard, B. S. (1991) *Systems Engineering Management*, Wiley, New York.

Chase, W. P. (1974) *Management of Systems Engineering*. Krieger, Malabar, Florida.

Hammer, W. (1972) *Handbook of System and Product Safety*. Prentice-Hall, Englewood Cliffs, N.J.

CHAPTER X

SUMMARY

REFERENCES

# 2
# Types of Test and Evaluation

*The first step toward improvement is to face the facts*
Oliver Wendell Holmes

By its nature T & E is inextricably linked to risk. The ultimate objective of T & E is to assess risks, viz., to identify the risks in a program and to confirm that risks have been reduced to levels that are acceptable to the program's sponsors and to the system's or product's users. Any major development, manufacturing or construction program involves a variety of risks that some costs will exceed budget, that some schedule delays will be encountered, or that some specifications or capabilities being sought will not be achieved – any of which might equate to a serious loss for the program, such as financial loss and a loss of reputation. The challenge for the program managers is to

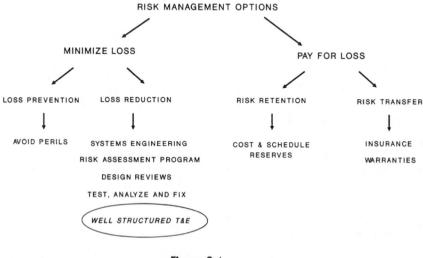

**Figure 2.1**
Treatment of risks

continually try to mitigate these risks. As shown in Figure 2.1, some risks are typically dealt with simply by accepting that a certain amount of loss is unavoidable, and by building in reserves to pay for that loss or buying insurance to cover it. Other risks are addressed by aggressive efforts to avoid the losses in the first place. When it comes to dealing with the losses that can neither be readily accepted nor totally prevented, a systematic, well engineered T & E program is essential.

## 2.1   TYPES OF PROGRAM

The types of risk vary considerably with the type of program. A program to design a household coffee maker with a built-in timer to activate it at a certain hour each weekday morning would likely not involve any substantial technological risk. Nor would there be a significant risk that a production line could not be established to produce quality coffee makers economically. However, there would be a marketing risk – a risk of not being able to capture enough of the sales market to make a profit or even to pay for the investments made.

The construction of an automobile bridge over a river is another venture where there would be negligible technological risks. Unlike the case of the coffee maker, there would not be anything like a marketing risk. However, there would be a substantial production risk. That is, there would be a risk that production type problems (e.g., labor costs, material delivery schedules, and delays due to weather) would jeopardize meeting the scheduling and cost constraints of the contract. Depending on how the contract were written, missing those requirements would likely incur financial penalties.

There is a third category of program: those that involve both a low number of products and a high level of technology. Space satellites, military combat systems, and the National Air Traffic Control System are examples. These programs involve significant development risks, because of the use of state-of-the-art technology; production risks because of the inability to amortize costs with so few products or systems; operational risks, because they operate in unique environments with dedicated logistics support infrastructures; and sometimes disposal risks, because they use relatively new materials for which the long-term environmental impact is not fully understood and for which safe and efficient disposal techniques have not yet been fully developed. Figure 2.2 correlates the types of risk that are predominant in the three categories of programs discussed above.

| | Low cost and large production | High cost & small production | |
|---|---|---|---|
| | | Low technology | High technology |
| **Development Risks:**<br>- cost<br>- schedule<br>- performance | | | ✓ |
| **Production Risks**<br>- cost | | ✓ | ✓ |
| **Marketing Risks**<br>- share of sales market<br>- return on investment | ✓ | | |
| **In-service Risks**<br>- operability<br>- reliability | | | ✓ |
| **Disposal Risks**<br>- costs<br>- safety<br>- environmental | | | ✓ |

**Figure 2.2**
Risks versus programs

## 2.2   TYPES OF T & E

Historically, the most common categorization of T & E was based on the levels of assembly of the hardware-intensive system:

- materials testing

- parts testing

- component testing

- subsystem testing

- system testing

That categorization of T & E is still useful for hardware-intensive systems, for large construction projects, and for the production of high-cost/low-quantity systems such as ships, aircraft, and satellites. However, it is more useful today, in light of the growing complexity of systems and their dependence on internal advanced digital computers, to categorize T & E in terms of the types of objective they are trying to achieve and the risks they are trying to assess, as categorized in Section 2.1.

(A) *Developmental T & E*:  T & E conducted to assist in the developmental process by determining the nature of risks, by confirming that adequate technical progress is being made, and to verify that corrections and improvements achieve desired results and do not themselves introduce new problems. Developmental T & E is an integral part of the systems engineering performed during Research and Development (R & D) programs.

(B) *Production T & E*:  T & E conducted to confirm that systems coming off the production line or items completing construction meet the production specifications and are ready to be placed in operation.

(C) *Marketing T & E*:  T & E conducted to confirm that a product will achieve the anticipated and minimum necessary sales levels. Marketing T & E is common in commercial products intended for high-rate production and low-cost sale.

(D) *Operational T & E*:  T & E conducted to project or to confirm that the product or system will have the required 'operational' capabilities and characteristics when it is placed in service. This type of T & E measures the acceptability to the user.

(E) *In-service T & E*:  T & E conducted to assesses the continued availability of the system and to project its state of readiness in the near future.

(F) *Disposal T & E*:  T & E conducted to verify that disposal of a product or system is complete and that the final conditions of the remnants or debris are acceptable.

While these different types of T & E correspond somewhat to the classic phases of a system's life cycle, it should not be implied that each type of T & E is exclusively conducted during such a phase. On the contrary, both the risks and the associated T & E freely defy such boundaries. Figure 2.3 shows some of each of the four types of T & E in development programs being conducted in almost every phase of a notional development program; even that figure is

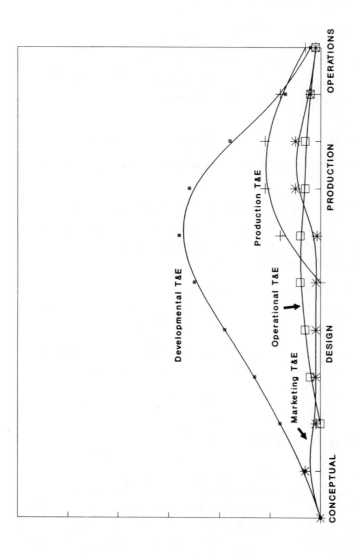

**Figure 2.3**
Types of T & E during a development program

oversimplified. The following discusses each of the major types of T & E in more detail.

## 2.2.1  Developmental T & E

Developmental T & E encompasses many different types of tests, analyses, demonstrations and inspections that are necessary during initial system development and fielding, as well as during the development of upgrades. It includes piece-part testing, component, subsystem and system level T & E, conducted at laboratories, manufacturing plants, integration sites, and operational sites. It includes:

- experiments to characterize performance,

- verification tests to assure technical progress is being made and requirements are being met,

- reliability growth tests to identify candidate design improvements,

- environmental tests (temperature, humidity, vibration, shock, radiation, corrosion, etc.) to confirm ability to perform in the extreme conditions in which the system may be operated,

- computer software stress testing, and

- life tests to identify obstacles to meeting shelf life or operational life requirements.

Any organization doing systems engineering will also be conducting some developmental T & E as part of the build–test–fix–retest process.

## 2.2.2  Production T & E

Production T & E verifies that a system will meet or has met its production specifications. When production starts, a new set of risks and uncertainties is introduced because there inevitably are differences in material, assembly techniques, parts, and computer software from what was tested previously. Production T & E verifies that the performance that was known to have been achieved is not lost or has not deteriorated in the production process. Production T & E includes environmental tests of prototypes, first article tests, sampling of production units, and some integration tests whereby production units are operated with or in other systems with which they must interact when they are placed in service.

### 2.2.3  Marketing T & E

In order to estimate to what extent a commercial product will sell, marketing T & E (also referred to as 'test marketing') is sometimes conducted on small, representative populations – before full-scale production and distribution are initiated. A common example of this is when a 'fast food' chain tries out a new entré in selected cities. Like all of the types of T & E, marketing T & E is difficult because the T & E process and procedures themselves may contain risks. In the early 1980s, the Coca-Cola company abandoned its decades-old successful formula in favor of a new flavor that had shown success in marketing T & E. Coca-Cola not only did not achieve the anticipated sales growth, but actually dropped in sales. Within two years, Coca-Cola returned to its old formula with a huge new sales campaign, dubbing it 'Classic Coke'.

### 2.2.4  Operational T & E

Operational T & E is conducted to assess to what extent a system will have or has achieved the required 'operational' characteristics and capabilities in service. Until the 1970s, this type of T & E was not recognized as a separate category. Full-scale system T & E routinely conducted as part of developmental T & E was relied upon to uncover all system-level problems before production. However, as the complexity of system designs grew, the T & E procedures became more technically specialized. Developmental T & E in the earlier phases of programs became more extensive, at the expense of later T & E, which became much more costly to conduct. Cost and scheduling constraints also created pressures to limit full-scale developmental T & E. The results were many examples of major programs delivering systems that did not perform as anticipated or required. The need to and value of collecting 'operational data' after a system was placed in service was recognized. However, after the production line has been turned on, there is little opportunity to make significant design changes because of the costs involved. In 1972, the US Defense Department defined Operational T & E as separate and distinct from developmental T & E, and made it policy that operational T & E be conducted *before* each major system proceeds into full production. Soon thereafter, similar policies were adopted for all development programs that the US federal government manages, particularly those in military, energy, space, and transportation systems programs. Operational T & E can also be seen in most complex commercial products, and even in some simple ones. For example, as part of the development of a new toy, Fisher-Price, Inc. subjects it to a series of use and abuse tests, having groups of children play with the toys for an extended period of time, while being observed behind a one-way mirror by Fisher-Price engineers and quality assurance personnel (Aronson 1978).

## 2.2.5   In-service T & E

This T & E encompasses the measurement, by the users, of the performance of systems and products while they are in service, well after the umbilical to the developers has been cut. Our ever-growing dependence on networks of complex systems and products in everyday life has necessitated not only highly reliable systems but good built-in T & E techniques for monitoring the status of performance of the systems and for identifying at what point the continued operation of the system will be in jeopardy.

## 2.2.6   Disposal T & E

In the last decade, a growing awareness of the depletion of natural resources as well as the damage to mankind that some of the new products of this century's technology are having on our environment have spawned many efforts aimed at carefully and safely disposing of certain materials. Disposal engineering is just now starting to become a profitable business in some product lines. Little is yet written about disposal T & E, but it is and will remain for some time an integral and indistinguishable part of the related engineering efforts.

## 2.3   SUMMARY

T & E can no longer be considered simply a part of – or constrained by – the levels of assembly of a new system or product. Government and corporate culture now values T & E not only as a product development tool, but also as a way of measuring user satisfaction. The emergence of operational T & E as a separate type is evidence of radical changes in systems architecture as well as in the development processes for new systems. It is also evidence of a new type of risk facing developers – the serious risk that the customer may not be satisfied with the product, even if it reflects better quality and a cheaper price than the alternatives. The need to satisfy better informed and more demanding customers is forcing T & E to take on dramatically new roles in system developments.

It is about the new roles of T & E in the development of complex systems and products that the remainder of this book is written. Developmental, production, marketing and operational T & E in today's development programs will be analyzed and synthesized. In-service T & E and disposal T & E, although listed above for completeness, are not treated to any significant extent, since they are emerging fields for which little experience has been captured or documented. Moreover, testing associated with basic research is not included because it does not entail the evaluation of results against objectives. As rocket scientist Verner von Braun said, 'basic research is what I'm doing when I don't know what I'm doing.'

## REFERENCE

Aronson, R. B. (1978) Toy Safety Isn't Child's Play. *Machine Design*, December 7.

# 3

# The Evolution of Recent T & E Practices

*Experience is a hard teacher because she gives the test first, the lesson afterwards*
Vernon Law

T & E has always been a part of design and engineering. It is an integral part of the iterative development process (build–test–fix–retest), and the evaluation of test results has always played a role in deciding when to market a new consumer product or to place a new system in service. But in the past several decades, T & E has become distinct and formalized. Common practices have evolved for the planning, execution, and reporting of T & E. In addition, the validity of the T & E process itself has been well studied, analyzed, challenged, and even used for marketing purposes. In this chapter, we will review the two parallel tracks along which T & E has evolved: in the commercial arena and in government programs.

## 3.1  BACKGROUND

In the 1950s, technology started putting more and more complex items in the hands of consumers. Automobiles, televisions, and telephones became common household items. Although more people began to own such products, they were still major investments for the typical household. Consumer activism was born from design problems with products such as these. Governments became involved in protecting consumers. For example, in 1962, amendments to the Food, Drug, and Cosmetic Law required pharmaceutical companies to prove the effectiveness of new drugs before marketing them. About this time, in the 1960s, we crossed a threshold in technology: we built things we could not fully test. The digital revolution yielded both many government developed systems and consumer products that could not be completely tested before they were given to the customer. In the 1980s, complex products became not just a luxury for most households, but a practical necessity. Here again we passed another threshold: we found we could not build products without human error. Governments and consumer activists insisted on good T & E. In the 1990s, for many items, the technology of testing lagged further behind the design capabilities. We launched space shuttles, introduced high-speed telephone switching networks, and deployed complex military systems that could not be fully tested even over the course of their lifetimes, much less while they were under development. We came to rely on simulating what we could not actually test.

In the 1970s, the US federal government recognized that technology and the associated risks warranted the establishment of policies for new systems being developed by and for use by the government. The increased use of computer technology was the major catalyst that forced the recognition that the trend towards increasing complexity in system design would continue to accelerate in the future. The United States Office of Management and Budget issued a landmark policy that applies to all major development programs being managed by federal government agencies. Regarding T & E, it says (USOMB 1976):

> Full production may be approved when system performance has been satisfactorily tested, independent of the agency development and user organizations, and evaluated in an environment that assures demonstration in expected operational conditions.

That policy has several key phrases which reflect where the US government felt there had been shortcomings in T & E. First, full system performance must be tested *BEFORE* production is started. Secondly, the system must be evaluated in environments that are as close as practical to the eventual conditions under which the system will be operated and maintained. Thirdly, T & E must have some visible independence from the influence of the developers as well as from the users of the system.

This policy document targeted in particular US government funded programs in defense, energy, space, and transportation. Each of the government agencies with programs in those areas published and refined its own implementing policies during the course of the two decades that followed. In T & E, the Defense Department clearly has the most formalized policies and procedures.

## 3.2   T & E IN THE COMMERCIAL ARENA

One could barely imagine a few decades ago how dramatic the change in the development of commercial products would be as they came to compete not just in local, stable market-places but also within the quickly growing global economy. The companies and product lines that survive today are those that are able to edge out their competition by being the first to provide a quality product at an attractive cost. The three operative and interacting objectives are: (1) to get the product to the market place as fast as possible, (2) to design and produce it as efficiently as possible, and (3) to maximize customer satisfaction. Achieving these objectives is particularly challenging at a time when systems and products are becoming significantly more complex, are multi-purpose and multi-functional, and must be more interoperable with other systems. Reductions in development times are being achieved by greater use of computer aided engineering, overlap of some development phases allowed by well engineered T & E conducted throughout the program, and getting the involvement of everyone from the intended users to the parts suppliers. Costs are being reduced by centralizing management and assets, and by substituting modeling and simulation for some full-scale prototyping and T & E. Customer satisfaction is being achieved by more operational T & E earlier in the program. These changes have resulted in three distinct characteristics of the T & E of commercial products that are expanding: the first is in the complexity of T & E techniques, the second is in user involvement in T & E, and the third is in the organizational identity of T & E.

## 3.2.1  Complexity of T & E

The growing complexity of T & E is apparent from the facilities that are used during design and development of today's systems and products. The photos in

**Figure 3.1**
Crosswinds testing of a GE aircraft engine

**Figure 3.2**
In-flight hail ingestion testing of an aircraft engine

Figures 3.1 and 3.2 show some of the testing conducted by General Electric (GE) in its aircraft engines development work. The first shows a crosswinds test site near Peebles, Ohio, and the second shows in-flight ingestion testing of a CFM56-3 engine produced by CFM International, a joint company of GE of the US and Snecma of France.

Through much team-oriented engineering and the extensive use of computer aided design and manufacturing tools, GE was able to reduce its development process for large aircraft engines in the early 1990s from 42 months to an average of 24 months. One of the key innovations that have allowed GE to have confidence in the engine performance while greatly reducing the development cycle time has been the use of 'flying test beds'. As early as practical, GE installs production representative engines in actual aircraft for in-flight T & E. Among other things, this T & E has allowed GE to:

- wring out the entire propulsion system early

- optimize the design characteristics

- evaluate the transient capability

- evaluate component performance in flight

- identify aero-mechanical problems early

- help the airframe manufacturer better plan and conduct his program to demonstrate compliance with US Federal Aviation Administration standards.

### Case example: Ford Motor Company

Companies are recognizing the growing investment they have in their T & E facilities and their T & E people. In 1995, the Ford Motor Company consolidated its assets into its new Global Test Operations (GTO) organization to promote more sharing among its product divisions, and thereby reduce costs. At that time, the GTO consisted of 5400 people, with facilities in the United States, Germany, England, Belgium, Mexico, Brazil, and Australia. As part of this consolidation, the GTO took over centralized management of test assets to reduce costs. Through the increased use of computer aided engineering (CAE) tools, improved knowledge about crash damage, and careful scrubbing of T & E program assumptions, Ford reduced the number of different crash models it uses from almost 140 to a little over 90, and the costs of the prototypes have been reduced from $40 million to $35 million. The company also invested in better test laboratory facilities such as Road Load Simulation equipment in its Dearborne, Michigan, plant pictured in Figure 3.3. Such rigs simulate the road profile inputs to the vehicle to assess durability, and have enabled Ford to greatly reduce the development cycle. Aside from increased use of CAE, Ford is reducing its T & E costs through better, earlier planning, lowering the number of alternative tests to be conducted, and earlier involvement by the parts suppliers. Figure 3.4 shows Ford's overall vision of lowering the costs of full vehicle-level testing.

**Figure 3.3**
One of Ford's road load simulation rigs

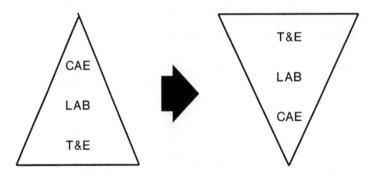

**Figure 3.4**
Lowering the cost of vehicle testing at Ford

The growing complexity of systems and the increasing criticality of sophisticated computer software have driven T & E to become a specialized discipline. One indicator of this is the growth of independent T & E laboratories. Well over 7000 organizations in the United States list T & E as their primary business, and another 9000 list it as their secondary. T & E laboratories provide trained people and dedicated facilities that engineering and manufacturing organizations could not easily build or maintain. The volume and repetitiveness of their work allows what might be complex work for another company to be handled routinely by a lab, with reduced costs and quick turn-around times. One of the well known companies which does such independent laboratory testing is Wyle Laboratories, Inc., headquartered in California. Established in 1949, Wyle has over 10 nationwide facilities on over 600 acres. Its facilities are well equipped, offering a wide range of environmental, atmospheric, and electromagnetic testing capabilities. Such laboratories provide not just specialized expertise and facilities to their customers, but also an unbiased view. They accept jobs that range from little more than conducting prescribed test procedures to actually engineering the test program and evaluating the results. After the space shuttle CHALLENGER accident, Wyle was tasked by NASA to be involved in the accident investigation. Today, the company operates solid rocket motor test facilities at the NASA Marshall Space Flight Center in Huntsville, Alabama (see Figure 3.5), which have been used to qualify the solid booster joint redesigned after the accident.

An example of the quick growth of independent testing laboratories is the story of the Kansas Analytical Laboratory, whose founders saw in 1991 a market need for quick, inexpensive blood analyses, primarily for health insurance underwriters. Within four years, it became the world's largest blood laboratory, handling a business of over $200 million per year.

Another indicator of the growing importance of good T & E programs is that T & E people now are involved early in the development program, even before system components are available for testing, so they can influence or at least

**Figure 3.5**
NASA solid rocket motor test facility operated by Wyle Laboratories

better understand the design trade-offs and engineering decisions being made. In a typical development process at Ford Motor Company, their Global Test Operations group becomes intimately involved in the program after the Program Decision, the first major milestone in the program.

AT & T's Bell Laboratories' process for the development of their Definity PBX System offers unique challenges. Definity is a communication system serving large business and government customers, providing voice, data, and video communication with routing, billing, networking and queuing capabilities. It serves 16,000 customers in over 100 countries, and an average of 20,000 users making 100,000 telephone calls an hour. Bell Labs use an 'incremental development and test' approach, whereby each update of the system progresses from defining the architecture through requirements setting, design, unit test and system test. However, there is overlap between the cycles of the updates, allowing individual new capabilities to get to the market-place earlier, and allowing earlier and continuous customer feedback. Figure 3.6 depicts this approach. Since the customer base is so large and diverse, yet so important to the success of the program, Bell Labs rely on computer based 'operational profiles' to simulate the customer part of the environment during T & E. An Automatic Test Generation Tool is used to randomly sample the system configuration, typical user profiles, and loading conditions during T & E. Given that the test planners must select from an almost infinite number of tests that could possibly be run on a system, Bell Labs allow the test conductors time to 'freelance' after they have completed the preplanned tests.

### 3.2.2   User Involvement in T & E

Throughout the development of new complex systems, formal involvement by the customers and prospective users is becoming commonplace. No longer a luxury, users are being involved in developmental and operational T & E to limit the risk that the end-product might not meet their needs and expectations. As much as there are great scheduling pressures to introduce new commercial

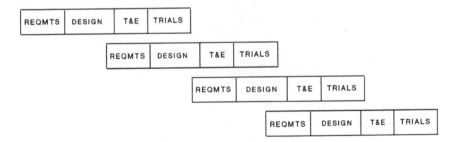

**Figure 3.6**
Incremental development and test approach

software for personal computers as fast as possible, no company today would do so without a 'beta test' – allowing a large cross-section of people representative of potential customers to actually use the software before its full release in order to uncover salient problems or weaknesses. Even the bigger 'fast food' chains like MacDonalds will do a several-month marketing T & E of planned changes to its menu before introducing them throughout the world. Probably the most visible driving factor in the schedule-intense development of Boeing's 777 aircraft was the close partnership that Boeing had with its initial customers such as United Air Lines. The early Extended Range Twin Engine Operations (ETOPS) certification of the first aircraft for United, described in Chapter 7, was achieved through close coordination between Boeing, United, General Electric (the engine manufacturer) and the certifying authority, the US Federal Aviation Administration.

---

### Case example: Fisher-Price toys

Fisher-Price, Inc., the world renowned manufacturer specializing in toys for youngsters up to 10 years of age has always had an impressive program for user involvement in T & E. Each toy made by the company is evaluated by a team consisting of representatives from product design, product development, manufacturing, quality control, and product safety. This team follows each toy from its inception to production. After manufacturing engineering has produced the first toys, using production material, tooling, and methods, they are subjected to a series of use and abuse tests, environmental, reliability and operational tests. After they are approved for production, toys are selected randomly off the first run of the production line for a repeat of the use and abuse testing. Safety is the focus of much of the testing at this point. Can the toy or any of its parts be swallowed? Can the user bite a hole in it? Are there any openings that might trap a user's fingers? But also of concern is whether or not the toy will sell. Before it goes into production, children are given samples of the new item, along with other toys, to play with under controlled laboratory conditions (Figure 3.7). Through one-way mirrors, trained observers note the children's reactions. They try to determine the 'play value' of each toy. They observe how quickly the child finds the toy, what he does with it, and how long it holds his interest (Aronson 1978).

---

## 3.2.3   Organizational Independence of T & E

The value of the organizational independence of at least some of the T & E people in a development program is becoming recognized. T & E has become so complicated and so specialized that there is widespread attention being paid to the validity of the methods and the objectivity of the results. Approval of electrical appliances by Underwriters Laboratories (UL) is a priceless marketing

**Figure 3.7**
Fisher-Price's 'Play Value' Laboratory

tool, because of their well recognized independent testing. The primary value of UL's approval of products is the perception by consumers that the testing is more objective than it otherwise would have been. Now over 100 years old, UL was founded by William Merrill, an electrical inspector who set up a small testing lab a few miles from Chicago's 1893 World's Columbian Exposition. The expo exhibited over 100,000 Edison incandescent light bulbs and highlighted the beginnings of electricity, for which safety was recognized as an issue. Today, UL operates almost 200 inspection centers in 63 countries. See Section 1.3 for photos of T & E at the UL facilities.

There are many other consumer oriented organizations, like Consumer's Union (CU), that do their own comparative testing of commercial products, not only to verify that the products perform as advertised, but also to evaluate how they will meet the needs and tastes of typical consumers. Consumer's Union is a non-profit organization established in 1936 to provide consumers, through its well known publication 'Consumer Reports', with information on goods and services. At their laboratories in New York, and at their Auto Test Center in Connecticut, CU tests selected brands and models and has *their* shoppers buy the items from stores, without disclosing their affiliation with CU. Under the scrutiny of CU engineers, products are evaluated for their performance, convenience, safety, and economy of operation. Many of the tests are those prescribed in industry or government standards, but frequently the CU engineers develop their own test standards and invent their own test equipment.

In 1993, the Australian Centre for Test and Evaluation was founded to create an organization focusing on the needs of T & E practitioners in the Asia Pacific region, but has actually attracted world attention. Affiliated with the University of South Australia in Adelaide, the Centre provides education in T & E, promotes technology transfer, and provides consulting services in planningand managing T & E programs. One of its early projects was to provide for independent T & E certification of gambling machines in South Australia.

In the increasingly challenging area of computer software development, the T & E programs are being used to provide vital measures of the maturity of new software during their development. Independent Verification and Validation (IV & V) of new software has become common practice. After the accident with the Space Shuttle CHALLENGER in 1985, the software programs in the shuttle were evaluated as part of the investigations. They were found to be of very high quality. International Business Machines (IBM), the contractor primarily responsible for the software, attributed part of that high quality to the fact that the software testers in their organization were kept separate from the design team (Joyce 1989). Nevertheless, lesser degrees of organizational independence have worked well on other programs. Microsoft Corporation, whose market is primarily individual Personal Computer users, makes the testers' ability to pace the delivery of new software to the market co-equal to that of the program manager. At Texas Instruments, on the other hand, whose primary market is in multi-platform, multi-database, multi-functional environments, having the testers working as part of a team under the direction of the program manager on each program has also proven successful.

---

### Case example: Nevada Automotive T & E Center

The Nevada Automotive Test Center has been in operation for more than four decades as an independent T & E facility, having completed programs with more than 1000 vehicle systems and millions of components for commercial, military, and public utility applications. The Center has more than 3000 miles of test courses that provide accurate simulations of commercial and defense related operating environments in over 100 countries world-wide. Its work spans the spectrum from structural analysis to prototype fabrication to production hardware certification. It is well supported by five test chambers for environmental, fungus, and electro-magnetic compatibility testing; sophisticated data acquisition and analysis capabilities; well equipped instrumentation and calibration facilities; and finite-element analysis, computer-aided design, and computer modeling and simulation facilities. It has experience in T & E of the latest cooling system requirements, collision warning and avoidance, antilock breaking, traction control, and both active and semi-active suspension systems. In addition, NATC does extensive testing of tires, and has developed a laser proofing system to augment other non-destructive tire-testing techniques. The photo in Figure 3.8 shows a four-tractor triple-vehicle combination, under remote control, providing pavement loading for purposes of evaluating new performance-related specifications for hot-asphalt mixes. This WesTrack project is being conducted by NATC and six other organizations under contract to the US Federal Highway Administration.

## 3.3   US DEFENSE T & E POLICY

The wholesale adoption by the US Department of Defense of strong T & E policies in 1972 was part of a dramatic change in the acquisition process. Even today, those policies remain unique, not only among US government and commercial development programs, but even compared to the government procurement policies of any other country. Nevertheless they represent an excellent set of comprehensive policies that provide a useful baseline against which, for academic purposes, one can compare other T & E policies. Those policies, although uniquely appropriate for the US military development programs, have been very visible and large scale – and have been quite successful.

**Figure 3.8**
NATC experimental road testing

## 3.3.1   The System Development Process

When US government policies started addressing the role T & E should play in each of its system development programs, the procurement of military systems was the first to come under scrutiny. Prior to that, T & E for military systems was treated solely as part of development and production engineering, conducted primarily by the contractor with some government witnessing. There were several reasons why that changed. First, military systems were costing significantly more than in the past, and were becoming a bigger part of the national budget like they were in the budgets of many countries. Second, technology in many areas was advancing more quickly than ever before, so that even during the 10 to 15 year course of a development program for a new system, there was uncertainty over whether unforeseen technology advances would necessitate substantial changes in the program even before the system was fielded. Third, the threats from other countries that the US government was trying to keep pace with (and stay ahead of, if possible) were themselves advancing rapidly. In other words, the arms race was on. And fourth, the policies that the US Defense Department had tried in the early 1960s to deal with this new environment through the 'total package procurement' had failed, and radical changes were needed.

Initiated by the then Secretary of Defense Robert MacNamara in the early 1960s, Total Package Procurement (TPP) involved – for major new systems – selecting early in the program a single prime contractor who would bring the system through both development and production. The selection of the contractor had to be based more on analytical studies than on proven system performance, since that contractor was selected so early in the program. There was somewhat of an 'arms-length' relationship between the government and the contractor: the specification he was given was primarily a performance-oriented one, in contrast to the detailed specifications that the Defense Department had previously used. The government attempted to minimize changes to those performance specifications because of the difficulty in determining the disruption it would cause the contractor and therefore the financial adjustment to which the contractor would be entitled. For these reasons, the changes that the government wanted to make that were not absolutely necessary were not made until after the production units were delivered from the contractor. The government had little influence in major design reviews and conducted no significant testing itself until production systems were delivered. Essentially, the contractor decided when and how to make the transition from R & D to production. This was all done to give industry maximum flexibility in pursuing innovative approaches to meeting system requirements, while reducing the sharing of technical risk between the government and the contractor. It would be the government's prerogative not to accept production systems from the contractor if they did not meet system performance requirements. So from the government's perspective, the most important T & E was that done after the delivery of production systems; the government had little influence on the testing that was done earlier or what the contractor did with what he learned from his testing.

The anticipated benefits from Total Package Procurement did not accrue as expected. Influences beyond the control of the program team – the changing threat, annually changing Defense Department budgets, rapidly advancing technology in many areas – made it impractical for the government to treat system performance requirements as stable for the five or more years it would take a contractor to get to the point of delivering production systems. And the uncertain nature of military systems R & D itself made the entire proposition too risky for some contractors.

There were some notable successes, but also some notable failures, in Total Package Procurement. But even the successes had their shortcomings. There were usually more changes and updates required to be made to production systems than were ever anticipated. The systems as delivered were somewhat outdated in many ways and had to undergo major upgrades right after delivery. The costs were significant, and the adverse publicity about the Military Services' new systems being 'obsolete' was damaging. In 1970, the Defense Department replaced the Total Package Procurement policies with the Milestone Procurement Process (MPP).

By contrast, Milestone Procurement and the related T & E policies have given T & E a more important role in the development process than it has ever had. The Milestone decision points that are inserted in each program are opportunities for all parties to review the latest forecast of the threat, latest state-of-the-art advances in technology, the defense budget – *and the T & E results to date* – to see whether continuing to proceed according to the current program plan is appropriate. Figure 3.9 shows a nominal comparison of Total Package Procurement with Milestone Procurement in terms of funding committed over the course of time. The government commits much more funding earlier and at a faster rate under TPP. If something were to force cancellation of the program, the government's financial exposure would be substantial. Milestone procurement, on the other hand, provides for a slower, more conservative commitment of funding, and several well placed opportunities to reset the program's cost, scheduling or performance objectives. This chart implies that eventually both cost/time lines converge. This is probably not accurate. As said previously, because of the upgrade programs that had to follow many TPP system deliveries, and the fact that the upgrade had to be made on many production items in contrast to updating only a few prototypes during the course of a Milestone Procurement Program's R & D, the final costs of TPP programs are probably greater.

### 3.3.2  Need for Change

Because of the responsibilities of Congress for the federal budget, several of its committees track and influence the Defense Department's acquisition policies

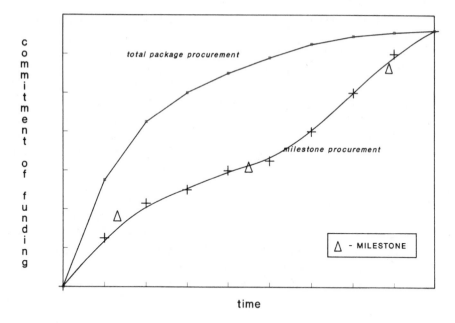

**Figure 3.9**
Total Package Procurement vs. Milestone Procurement

and processes. In the early 1970s, T & E became a regular topic covered in such discussions. There were frequent criticisms in the Committees' reports that operational T & E was not started early enough in military development programs. The reports said that the suitability of a system to be operated and maintained in a combat environment needed to be determined prior to production, not after the system is placed in service. The committees also complained that what T & E had been conducted in operational environments did not appear to be sufficiently independent of the influence of the developer, so the objectivity of the results was suspect.

The US General Accounting Office (GAO), which conducts independent audits of government programs for Congress, undertook a series of reviews of Defense Department development programs. Based on those reviews, the GAO published a report (USGAO 1972) on T & E in those programs. It concluded that the establishment of test objectives in development programs was adequate. However, the test plans were judged inadequate because system level operational T & E was not being conducted prior to the time that approval for production was planned. The GAO concluded that T & E was not timely or effective. Test reports were adequate, but their value was diminished because test completion dates were set on a relatively inflexible basis. Schedules were unduly optimistic and success oriented. When T & E disclosed deficiencies after system production was started, the repairs would be more expensive than if they had been made prior to production. The GAO also concluded the information given to the top-level decision makers in the development process was of 'diluted value'.

In 1970, President Richard Nixon convened a Blue Ribbon Defense Panel of government and industry experts to review the Defense Department's systems development process. That panel looked at T & E and concluded that the developmental testing conducted to determine how well systems met technical requirements and contract specifications 'appears to be fully well understood and faithfully executed'. The panel concluded that this type of testing did not appear to be a problem area. But the panel went further to make the observation that it would be 'extremely useful to replace or support critical assumptions and educated guesses [about system capabilities] with quantitative data obtained from realistic and relevant operational testing'. The writing about military systems T & E at this time began to make a distinction between the technical, developmental T & E that appeared to work well and operational T & E that would have uncovered, before fielding, problems with an operational impact that developmental engineering and T & E was not uncovering. As the thinking about 'operational T & E' evolved, its definition became very distinct from developmental T & E.

### 3.3.3   The T & E Policies

On January 19, 1973, the US Defense Department issued its first department-wide policy directive on T & E in system development programs (USDoD 1973) – policies which essentially have remained the same since that time. The first major policy is that operational T & E is to be conducted prior to a program proceeding into full-scale production to verify that new systems are *operationally effective* and *operationally suitable*. This policy institutionalized operational T & E throughout the Defense Department, forevermore distinguishing it from developmental T & E. Figure 3.10 highlights the major

| DEVELOPMENTAL T&E | OPERATIONAL T&E |
|---|---|
| controlled environment | free-play environment |
| use of skilled operators | wide range of typical users |
| data collection paramount | realism of operating conditions paramount |
| number of variables minimized | limiting variables not a high priority |
| focus on technical parameters | focus on operational measures |

**Figure 3.10**
Differences between developmental and operational T & E

differences between the focus of developmental T & E and operational T & E in the US Defense Department. An appreciation of the types of system design deficiencies that the US Defense Department is trying to prevent from getting to the operational forces can be gleaned from the definitions (USDoD 1973) of the terms operational effectiveness and suitability:

Operational *effectiveness* is 'the overall degree of mission accomplishment of a system when used by representative personnel in the environment planned or expected (e.g., natural, electronic, threat, etc.) for operational employment of the system considering organization, doctrine, tactics, survivability, vulnerability and threat.' In simpler words, it is the capability of the system to perform its mission.

Operational *suitability* is 'the degree to which a system can be placed satisfactorily in field use, with consideration given to availability, compatibility, transportability, interoperability, reliability, wartime usage rates, maintainability, safety, human factors, manpower supportability, logistic supportability, natural environmental effects and impacts, documentation, and training requirements.' In other words, it is the ability of the user to operate and maintain the system in its environment over the long term.

The Defense Department has set rigid requirements for what qualifies as operational T & E:

1.  Operational T & E is to be conducted in the intended operational environment, under realistic conditions. If that environment is a Navy ship in heavy seas, an Air Force tactical bomber flying in rough weather, or an Army tank tackling difficult dessert terrain, that is the environment in which the system is to be operationally tested.

2.  Operational T & E is to be conducted on a production-representative system. Experience has proven that many programs lose some effectiveness and suitability when transitioning from building R & D prototypes to building assembly line production units.

3.  Manning of the system during operational T & E must be by typical operators and maintainers with the same qualifications and training that would be expected of those who will operate the system while it is in service. People involved in the development program cannot participate in Operational T & E, lest their participation interfere with the evaluation of the typical crew's ability to operate the system in all modes and under all conditions, to adequately diagnose problems as they occur, and to properly correct those problems in a timely manner.

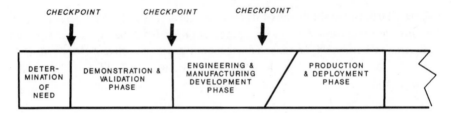

**Figure 3.11**
Notional US military development program

The second key T & E policy is that acceptable T & E results are to be a significant consideration at each program's milestone approval reviews. 'Try before buy' became a phrase often used to describe this T & E policy, and indeed it captures the essence of the change in the entire development process in the early 1970s. Programs are to be structured so that major milestone decisions are planned at points in the program where meaningful information, including T & E results, are available to the decision maker. Likewise, T & E events are to be planned so that they maximize the value of T & E results available at each decision point.

The major phases of a notional system development program are depicted in Figure 3.11. Prior to beginning the Demonstration and Validation Phase, the Engineering and Manufacturing Development Phase, and the Production Phase, there is a major checkpoint – a 'Milestone Decision' as the US Defense Department refers to it, and a 'Key Decision Point' as the US Department of Transportation calls it – at which a total review of the program and its objectives is conducted to decide whether it should proceed as planned, be revised, or even be cancelled. The most important milestone decision is the one that approves a program proceeding from R & D into full-rate production.

The third Defense Department T & E policy is that each Military Service will have an Operational T & E Agency (OTA) that will be responsible for planning and conducting operational T & E. The agency is to be organizationally separate and independent from development organizations and from the user organizations, the operational forces. They are chartered to provide an assessment of each system's operational effectiveness and suitability to the appropriate Military Service Chief. The agencies are:

- the Air Force Operational Test and Evaluation Center (AFOTEC), head-quartered in Albuquerque, New Mexico;

- the Army Operational Test and Evaluation Command (OPTEC), head-quartered in Alexandria, Virginia;

- the Navy's Operational Test and Evaluation Force (OPTEVFOR), with headquarters in Norfolk, Virginia;

- the Marine Corps Operational Test and Evaluation Activity (MCOTEA), Quantico, Virginia.

The Office of Secretary of Defense has a Director of Operational T & E (DOT & E) who reports directly to the Deputy Secretary of Defense. DOT & E's office was established in 1983 in response to changes in Public Law written by Congress. DOT & E and his staff oversee each of the Services' implementations of the Defense Department's policies regarding operational T & E. Also, by law, he must report to Congress his assessment of T & E in each major acquisition program before the program enters full-scale production. He assesses both the adequacy of the operational T & E results as well as the validity of the operational T & E program for each such system.

The fourth key T & E policy is that top-level planning for each T & E program is to be documented in a T & E Master Plan (TEMP). For Defense Department programs, the TEMP is an executive level document, co-signed by the developing organization, the operational T & E agency, and the organization which sets the system requirements on behalf of the user. For major programs, DOT & E and the Secretary of Defense's other T & E office, the Director of Test, Systems Engineering and Evaluation, also review and approve the TEMP. The TEMP is discussed later in Chapter 5.

## 3.3.4   T & E During a Typical Development Program

During the conceptual phase of a military development program, rarely is any formal T & E conducted. However, if there are several significantly different types of systems that could meet the requirements, a technology demonstration may be conducted to help select the preferred alternative for the next phase. During theDemonstration and Validation Phase (sometimes referred to as 'advanced development'), T & E is conducted on components or subsystem prototypes to identify the major risks and to demonstrate feasible solutions. The Milestone Decision authority may request that an Early Operational Assessment be made by the Operational T & E agency. During this phase of the program, the design is firmed up to the point that the technical and operational performance requirements for the system can be finalized for the next phase. During the Engineering and Manufacturing Development phase (also called 'full scale development'), developmental T & E is conducted to identify significant design problems for correction and to assure that before production begins, engineering is reasonably complete. Prototypes or pre-production models are built, and operational T & E is conducted to evaluate the operational effectiveness and suitability of the system to verify that it meets minimum requirements, prior to full production. The last increment of operational T & E during production and deployment is a full-scale, independently conducted, event. Low-rate initial production is often authorized and started during this phase. During the Production and Deployment phase, development T & E is conducted to evaluate changes that were made to the system to correct previously identified deficiencies, to incorporate product improvements, to meet changes in the threat, and to improve reliability, maintainability and availability. Operational T & E is conducted to further evaluate operational effectiveness and suitability, to examine changes, to verify no loss in capability in going from the prototype version to the production version, and to help develop new tactics.

---

### Case example: The Mark 50 torpedo

In the early 1980s, the Defense Department established the requirement for a new torpedo, the 'Mark 50', to protect friendly Naval forces and commercial shipping from attack by enemy submarines, particularly fast, deep diving submarines. Figure 3.12 is a photo of the production version of the torpedo. The Demonstration and Validation phase of the of the program was conducted from 1979 to 1983, resulting in four competing conceptual proposals. Towards the end of the Demonstration and Validation Phase, 18 early prototype torpedoes were built. Laboratory test articles included propulsion, guidance and control, and warhead test sections. Over 100 runs of these torpedoes were made to characterize and improve the performance of the various subsystems. From 1986 to 1988, 18 Engineering Development Model torpedoes were built and tested, followed by 72 preproduction torpedoes from 1988 to 1990. These torpedoes were used in at-sea tests, launched against surface ships and submarinesin simulated attacks. Some of these tests were conducted by the Navy's Operational Test and Evaluation Force (OPTEVFOR) in 1988. The results of these tests had been intended to be the basis for approving low-rate initial production of about 340 torpedoes.

In early 1990, the program team conducted the Technical Evaluation (TECHEVAL) of the preproduction torpedoes, in preparation for the Operational Evaluation (OPEVAL). Torpedo sea runs were conducted at several Navy ranges with different environmental conditions. All of the torpedoes were prepared by the Naval Undersea Warfare Center, Keyport, Washington which later would become the Depot Maintenance Facility. Torpedoes were launched from several types of aircraft and surface ship, against a variety of representative targets, including actual US submarines in some simulated attacks.

In accordance with the 'try-before-buy' procedures, approval for full production (Milestone IIIB) was planned to be based on the results of this developmental and operational T & E. The reports of the Commander of OPTEVFOR to the Chief of Naval Operations contained not only a detailed assessment of the operational effectiveness and suitability of the torpedo, but also his recommendations regarding production. COMOPTEVFOR drew his conclusions from all of the operational T & E conducted during the program, and most importantly from the OPEVAL. OPEVAL is the Navy term for what is usually the last operational T & E event conducted during a program's Engineering and Manufacturing Development phase. In planning and conducting OPEVALs, OPTEVFOR tries to ensure T & E is done in a realistic operational environment. In the case of the Mark 50 torpedo, OPTEVFOR conducted all runs in as operationally realistic conditions as possible, while meeting the need for measuring accuracy in reconstructing the torpedo-to-target encounter. It evaluated the entire stockpile-to-target sequence for the torpedoes. Torpedo inspections and turn-around maintenance were conducted as they would be when the torpedo was placed in full service, and were evaluated at the Intermediate Maintenance Facility in Keyport. The OPEVAL was successful. However, the Defense Department decided to scale back the planned inventory objectives of torpedoes because of the collapse of the Soviet Union. Production was authorized and funded for a limited number of additional torpedoes, bringing the total number to about 1000.

Usually the system used is a full scale prototype or a pilot production model closely resembling the planned production configuration. During OPEVALs, the system under test is operated and maintained by Navy personnel of the same rates that are planned for production systems (their people will have received training representative of planned production system training). They will have spare parts support to the level that is planned for typical Fleet conditions. And they will have technical manuals – perhaps not finalized – but usable by the system operators and maintainers. During OPEVAL, OPTEVFOR tries to have the system operate in all of its significant modes and under the full range of typical conditions in order to be able to thoroughly evaluate its operational effectiveness and suitability.

**Figure 3.12**
The Mark 50 torpedo (US Navy)

The end of the Engineering And Manufacturing Development (EMD) phase is where the total-system T & E is conducted that is most closely associated with try-before-buy. But this T & E is of course the culmination of earlier T & E at the system, subsystem and component level that forms the building blocks of the whole program.

A major challenge for US Defense Department program managers implementing its T & E policies is a basic incongruity between those T & E policies and the budget process, both of which are mandated by Congress. By the T & E policies, programs should be flexible and milestone oriented, i.e., the program should proceed from one phase to the next based on the achievement of a (previously established) amount of technical progress. On the other hand, through the government-wide budgeting process, the Department operates on a fixed, calendar oriented basis. It must justify, as far as five years in advance, how much R & D funding it will need each year and when the program will move from using R & D funding to procurement funding. And the process makes it very difficult to move funding from one fiscal year to another, from one type of funding to another, or from one program to another.

### 3.3.5  Defense T & E in US Law

These basic policies have essentially remained the same since they were initiated in 1973. The changes that have been made since then are primarily additions to accommodate specific policies and procedures imposed by the US Congress and invoked in Public Law. In 1974, the requirement was imposed that in the then-annual Presidential Budget Submission, the Defense Department must describe the quantitative T & E results for major system development programs for which R & D or procurement funding was requested in that budget (US Code 1974). In 1983, Congress required that the Department establish an office of the Director of Operational T & E (DOT & E), reporting directly to the Secretary of Defense and the Deputy Secretary of Defense (US Code 1983). Since the early 1970s, T & E policy and oversight for both developmental and operational T & E had been performed by the one T & E office under the Under Secretary of Defense for Acquisition. However, Congress felt that this office was not sufficiently separated organizationally from the sponsorship and advocacy of development programs. In that legislation, Congress also required that DOT & E approve in advance the detailed plans for operational T & E events in major programs, and that he provide a report to selected Congressional Committees before each major program proceeds beyond low-rate initial production. Policy and oversight of developmental T & E has remained with the Director of Test, Systems Engineering, and Evaluation, reporting to the Under Secretary of Defense for Acquisition.

In 1987, Congress added special requirements for full-scale 'live-fire' T & E to demonstrate lethality in major ordnance and munitions programs and to demonstrate adequate protection against vulnerability in 'covered' system programs such as those for aircraft, tanks and ships (US Code 1987). Live-fire testing in defense programs is described in Chapter 6, Section 6.9. In 1989, Congress again revised the law, this time to add a requirement that major programs may not proceed beyond low-rate initial production until at least one phase of operational T & E has been conducted (US Code 1989).

## 3.3.6   Evolution of Operational T & E

The most significant impact that the US Defense T & E policies made was very formalized treatment of operational T & E as independent of the development process and as separate from developmental T & E. The attitude towards operational T & E and how 'independent' it needed to be has changed over time. In the 1970s, the Services were back-fitting operational T & E into programs already underway. So initially, operational T & E tended to have much involvement of the developer and user organizations. Due to pressure from Congress for more independence (as evidenced by Congress forcing the Defense Department to establish a separate office of the Director of Operational T & E), operational T & E of full-scale systems in the 1980s came to be more separate and distinct from developmental T & E and from even the appearance of undue influence of anyone involved in the development of systems. In the mid-1990s, due to the dramatic reductions in the defense budget, a 'middle ground' was reached. Emphasis was placed on more sharing of resources and data between developmental and operational T & E, more combined T & E events conducted earlier in the program, and smoother transitions from developmental to operational T & E – enabled by the operational T & E community's willingness to use more developmental T & E data in its analyses.

## 3.3.7   T & E in Other Countries' Military Procurement Programs

The United States is unique in the large numbers and diversity of military systems its industry produces, both for use in the US Military Services and for sales to other nations. For example, it is the only country that requires formal, independently conducted, operational T & E as a graduation exam before a program is allowed to proceed to full scale production. Other countries have much smaller organizations for military systems developments, with less internal-government engineering capability. Compared with the US, where the large companies producing military systems are diversified, also producing products for the commercial market-place, the companies in the military systems business in other countries are fewer and usually specialize solely in defense products. The United Kingdom and France essentially have nationalized defense industries. Compared with the US, other governments place much more reliance on the contractors to plan and conduct the T & E programs during development. Of course, there is government oversight of such T & E. The government defense departments of France and the United Kingdom have T & E agencies which manage the operation of their test ranges and help oversee the conduct of T & E during development programs. Design problems that are not uncovered during the development programs are corrected by the contractors when the systems are initially placed in operational service.

## 3.4   T & E IN OTHER US GOVERNMENT PROGRAMS

Development programs outside of the US Defense Department, in space, energy, and transportation, have T & E programs and policies tailored to their customers and system users, but nonetheless complying with the intent of the policy of the Office of Management and Budget that full production should not commence until operational T & E in the user's environment is conducted. The Federal Aviation Administration, for example, defines operational T & E to encompass: (1) OT & E Integration Testing of systems' end-to-end performance, (2) OT & E operational testing to verify operational effectiveness and suitability with user participation in the testing, and (3) OT & E Shakedown Testing, independent verification and validation of the system as part of the National Airspace System by the user organizations. An Office of Operational T & E Oversight monitors the implementation of this testing, approves selected detailed test plans, and conducts some of the operational T & E (USDoT 1992). The Coast Guard assigns responsibility for independent conduct and oversight of operational T & E in each major program on a case-by-case basis, usually to its Research and Development Center in Groton, Connecticut. The National Aeronautics and Space Administration does not distinguish between developmental and operational T & E in its space shuttle program. Essentially their T & E is a very intense methodological approach of testing one vehicle for one mission. That testing completes at the end of launch countdown.

## 3.5   SUMMARY

Both in commercial and government development programs, T & E is no longer being used only as a systems engineering tool. It is now used as a tool for assessing risk throughout the course of the program and for measuring customer satisfaction. New innovative ways of accurately measuring that customer satisfaction before a product is sent to full production and put into service are being aggressively pursued. T & E is becoming a visible part of the management and an important member of the integrated teaming arrangements that are characterizing today's development programs.

# REFERENCES

Aronson, R. B. (1978) Toy Safety Isn't Child's Play. *Machine Design*, December 7.

US Code (1974) Title 10, US Code, Section 139.

US Code (1983) Title 10, US Code, Section 138.

US Code (1987) Title 10, US Code, Sections 2366 and 2399.

US Code (1989) Title 10, US Code, Section 2400.

US Department of Defense (1973) *Test and Evaluation.* Directive 5000.3, January 19 (no longer in effect).

US Department of Defense (1991) *Defense Acquisition Management Policies and Procedures.* Directive 5000.2, February 23.

US Department of Transportation (1992) *FAA National Airspace System T & E Policy.* Federal Aviation Administration Order 1810.4B, October 22.

US General Accounting Office (1972) *The Importance of Test and Evaluation in the Acquisition Process for Major Weapon Systems.* Report #B163058, August 7.

United States Office of Management and Budget *Major System Acquisitions.* Circular A-109, (1976).

Joyce, E. J. (1989) Is Error-Free Software Achievable? *Datamation*, February 15.

# 4

# The Process: T & E Engineering

*Plan ahead — it wasn't raining when Noah built the Ark*
General Features Corporation

Today, successful systems engineering in a development program involves much more than the aggregate of engineering of the individual subsystems plus a phase of integration of the subsystems. So too, T & E engineering at the overall program level must be more than an aggregation of subsystem T & E engineering. T & E engineering must be thorough and disciplined. It must be well documented so that it can be readily restructured, if necessary, as the technical risks change during development. This chapter focuses on the different factors in T & E engineering that are requiring special increased attention today, particularly because of the growing size, complexity and interdependence of systems. Chapter 5 describes the basic framework of documentation critical to orchestrating and capturing this engineering.

Figure 4.1 shows the major activities within T & E engineering. The important aspects of planning and executing each will be discussed in this chapter. A more complete depiction of T & E engineering of complex systems today would include modeling and simulation; that is discussed in Chapter 6, Section 6.11.

## 4.1 REQUIREMENTS ANALYSIS

T & E personnel are not ordinarily thought of as having any responsibility for defining system performance requirements. Historically, they have often been

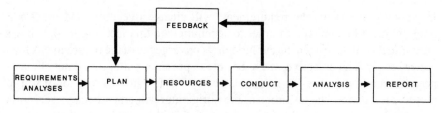

**Figure 4.1**
The T & E engineering process

cautioned that any actual or perceived participation by them in setting requirements could conflict with their objective evaluation of test results, and should therefore be avoided. However, during the full scale development of the complex systems of the 1970s and 1980s, many of the top level deficiencies identified during the T & E programs had their bases in inaccurately defined requirements rather than in poor system designs. To be adequate for use in the T & E program, system performance requirements need to be accurate, well documented, and well understood – leaving little need for interpretation by either the system designers or testers. Today's systems have three features that make the job of adequately defining the requirements a particularly challenging one:

- First, today's systems are multi-functional versus single-purpose. The customers and users are usually a wide variety of people with differing needs and preferences.

- Secondly, they are more interoperable with their environment than the systems of the past. A system that prints checks for bank customers must now accommodate a host of electronic handling equipment and systems that those checks will see from the time they are deposited in one bank until they are credited to the receiver's accounts and debited from the payer's accounts. A city's traffic lights are typically now part of an integrated traffic management system that can adapt to daily and hourly fluctuations due to rush hours, special sports events, accidents, and weather problems.

- Thirdly, the system requirements are dynamic. They will change during the course of a development program as customers' desires change, as public policies change, as other competing systems are introduced to the market-place, and as new technological breakthroughs occur.

There are three important considerations in analyzing system performance requirements for the purposes of the T & E program: their currency, their traceability to the user's needs, and the actual evaluation criteria.

## 4.1.1  The Currency of the Requirements

In complex programs, the requirements will change over time and the T & E program must be dynamic in order to accommodate this fact. Figure 4.2 shows what might well have been the evolutionary convergence of the requirements for a butter keeper. The objective was to be able to preserve butter. In all but climates near the equator, it was too cold to keep the butter outdoors, but too warm to preserve it indoors. The invention of the refrigerator helped with the preservation of most foods needing cold temperatures, but it was a bit too cold for butter. Eventually, the butter keeper was invented (... and designed and tested and produced) (Greenlee, 1992)! Consider the simple example of a common utensil, the fork. Thomas à Becket introduced the Italian fork to England in the 12th century. But it did not catch on immediately. For several centuries, civilized dining throughout Europe continued to depend on fingers and a pair of knives, one to hold the meat and the other to cut it. But a single-point blade does not hold a piece of meat securely. So fork experimentation continued, and forks finally became commonplace in the 18th century. There was no known formal T & E program for the fork or butter keeper. They are simple systems developed to satisfy simple – albeit evolving – requirements. But when the system is larger and more complex, there is a need for a formal T & E program. From the tragic earthquake in San Francisco in 1906 came a commitment to develop new techniques for designing buildings in that area with more structural tolerance for quakes, and new methods and standards for T & E. From pollution problems that became apparent in major industrial centers in the middle of the century came the need for pollution control techniques for automobile exhausts, as well as sophisticated T & E equipment to periodically measure the effectiveness of those systems in each automobile. As technology provides us with more and more complex systems, our T & E capabilities must keep pace.

Clearly the most challenging development program is one where the significant, dominant requirements are continuously evolving and changing at a rapid rate. This is particularly common in military system programs. The need for a new system or major system upgrade is usually based on the need to counter

**Figure 4.2**
The requirements for a butter keeper

specific threats, knowledge of which is always incomplete and changes frequently. System requirements are generally based on a combination of information about current threats, projected threats, and postulated threats, tempered of course by what can reasonably be afforded. During the typical seven- to ten year development program for a military system, the perception of the current, projected and postulated threats will change, and the officials responsible for setting the system requirements are likely to revise those requirements, even though such changes usually result in cost increases and scheduling delays. Military programs frequently push or anticipate using certain technologies, which may not materialize or turn out to be as adaptable to systems engineering as originally envisioned. This too adds to the risks and uncertainties of the development program being able to eventually meet the requirements. In this environment, the T & E program is used as a pacing item for the program, helping to ensure that too costly or unworkable approaches are not pursued and that premature designs are not fielded.

The requirements for military systems change because of the rapidly changing threat they are being designed to counter. But the requirements for 'host' or platform systems, such as commercial communications systems and military Command and Control systems change because the systems they service change periodically, and independently of each other and the host system. The incremental development and test approach used by AT & T in its Definity communications system, described in Chapter 3, section 3.2.1, is a good example.

Recently, computer aided tools have been developed to assist in defining, studying, and controlling – and describing to the customer – the ever evolving system requirements throughout the program. As mentioned previously, today's complex systems cannot be designed or tested without being mathematically modeled, and without their operations being simulated on computers. There are a variety of computer based tools available that can help build and maintain a multi-dimensional matrix that correlates system requirements with test requirements, and with program (cost and scheduling) changes.

## 4.1.2   Traceability to Users' Needs

It is important in establishing the T & E program requirements to keep a focus on the user. As the original objectives of the program are translated into performance requirements, and then possibly into a host of specifications, it is easy to lose full traceability to the original performance objectives. The people developing the detailed engineering specifications are sometimes significantly removed in time, physical location and organizational relationships from those who wrote the requirements. How easy it therefore is for the many people within the company and the subcontractors and suppliers to lose sight of, or even misinterpret, those requirements.

The user's environment and habits must be well understood and kept in focus at all levels of the T & E program. An *operational profile* should be developed which describes the length of time the system would be operated in various modes or conditions, the various environments in which it would be operated and maintained, and the expected spectrum of users. That profile should then be used to help optimize the design not only throughout the system's engineering but also during the failure modes and effects analyses, reliability demonstration tests, design-limit testing, life tests, as well as all of the developmental T & E and operational T & E events that will be conducted. Operational profiles are frequently modeled to allow for quick, relatively inexpensive excursions while designing and testing the various ever-increasingly complex usage profiles of today's systems. American Telephone and Telegraph (AT & T) treats the operational profile as an indispensable tool in every T & E program. It models the expected customer environment, including large organizations, individual users, multinational corporations, varied types of business, and purchases from catalogs. For military systems, operational profiles are mission-based models of system usage, for which a wide variety of complex combat scenarios can be conducted for both design and T & E purposes.

## 4.1.3  Appropriate Evaluation Criteria

At the beginning of a program, a number of performance parameters should be selected that represent the dominant system characteristics. They should be selected such that progress towards achieving them can be measured in T & E, and such that their achievement is a good indicator that the performance requirements will ultimately be met. Evaluation criteria should be developed for each parameter to enable relatively unambiguous assessments to be made of progress during the course of the program and of achievement of final system performance at the end.

System development has become so complex and so costly that T & E *must* be used to pace and help focus the engineering and design efforts of a program. Even early in the development program, before the ultimate technical approach has been selected, T & E in conjunction with modeling and simulation can help assess where the most promising solutions may be so the program can husband its limited resources. During development, the program's investment sponsors can be provided with interim T & E results to confirm that the approaches being pursued continue to be workable and to be the most promising. Like the design effort itself, the T & E program must be well engineered to accomplish this. Today, the importance of T & E at the *end* of development has been surpassed by the use of T & E as a tool for assessing progress *during* the program.

For T & E to confirm that the objectives have been met at the end of a development program appears conceptually simple. Rigor and discipline are usually applied to defining the end objective, normally in appropriate quantitative terms; e.g. an automobile with specified speed, acceleration characteristics, fuel economy and habitability features, that will capture a certain share of the new-car sales. Those quantitative terms then become the measurement criteria for the T & E program. But for the T & E program to objectively assess interim progress earlier in the program is much more difficult, especially for programs that push the state of the art. The operative word here is 'objectively'. In the past, the mostly subjective judgment of the developers was sufficient for progress assessments. However, as systems, subsystems, and components have become more complex, subjective judgment has proven not to be sufficient. And as programs have become more costly, and difficult to assess, developers have become challenged more by their resource sponsors – the government, a group of investors, the corporation's Board of Directors – and the extent of this objectivity has come to be suspected because of their advocacy of the program. In response, the sponsors have insisted on more precision and objectivity in assessing progress. They have insisted that a well structured and engineered T & E program be agreed to in advance for assessing progress.

In the mid-1970s, the US Defense Department instituted a very deliberate and highly visible procedure that significantly increased the formality required in the T & E programs for its systems. Early in a development program, key performance parameters are identified and 'threshold' values are assigned for use in T & E. The following discusses the parameters and the thresholds separately.

## Parameters, or MOPs

Parameters, or Measures of Performance (MOPs) as they are sometimes called, are keys to successful evaluation of the results of the T & E program and to substantiating the conclusions and recommendations made based on those results. (For operational T & E purposes, MOPs are further divided into Measures of Effectiveness (MOEs) and Measures of Suitability (MOSs).) Experience in US Defense Department programs demonstrates the pitfalls in selecting MOPs and suggests how to avoid them (Kass 1995). First, the interrelationship between parameters needs to be documented and understood. It is sometimes useful to categorize measures in a hierarchy such as primary measures, supporting measures, and diagnostic measures. Second, parameters should not be over-specified. Including mean, median, or mode values or confidence levels as part of the parameter or threshold is usually not appropriate. More often that not (except for measures of reliability and availability), they are part of the test procedures and should not be misunderstood to be part of the pass/fail criteria. Also, parameters are sometimes suggested which are actually requirements for data to be collected rather than performance measures in themselves. An example would be the number of messages sent by a communication system. Also listing test conditions should be done in the test plan rather than become part of the parameter. Characteristics that are not critical to system performance (e.g., size and weight in most cases) should not be MOPs. Features that bound the performance of the system but are not design risks that could jeopardize that performance, such as input power requirements or transportability, should not be MOPs.

*Threshold values*

Threshold values are understood in these programs to be the pass/fail criteria; if performance as measured in T & E falls short of even one of these thresholds, the program is in jeopardy of being delayed or perhaps even cancelled outright. To reinforce the importance of properly identifying these thresholds and adequately testing to them, the US Defense Department in 1991 began to refer to them as 'Minimum Acceptable Operational Performance Requirements'. Having so carefully defined such pass/fail criteria for the program, however, left the need for other less 'absolute' metrics. In addition to being an input to program investment decisions, T & E helps gain more insight into the system's capabilities and limitations, suggests ways to use and tactically deploy the system, and provides feedback to the government and industry engineering communities that will be useful in future system upgrades and the design of the next generation of systems. So in the mid-1990s, the Defense Department institutionalized the use of metrics in top-level program documentation which are not used in making programmatic (pass/fail) recommendations, but are used as benchmarks in assessing the system's general operational utility.

When the top-level T & E requirements are considered program pass/fail criteria as they are in US Defense Department programs, it is critically important to ensure that the T & E parameters are the appropriate ones. Even for the most complex system, they should be only the top 10 or 20 that are the dominant characteristics of the system's performance. And the threshold values associated with each parameter should be the correct ones. They should be the 'show-stoppers', reflecting a high level of performance. If this level is not achieved, the program should not be allowed to proceed. There are typically thousands more parameters of interest, but it is from these top-level ones that nearly all others flow. Figure 4.3 depicts a 'flow-down' relationship between these top-level criteria and those in lower-level documents.

Where possible, the top-level evaluation criteria should be traceable to the lower-level T & E documents throughout the program, so that when design trade-offs are made, achievement of the minimum required performance is not jeopardized. This is not to say that for a given parameter, the pass/fail value needs be the same through the hierarchy of program documents. On the contrary, that value should be more stringent in lower level documents, in order to allow

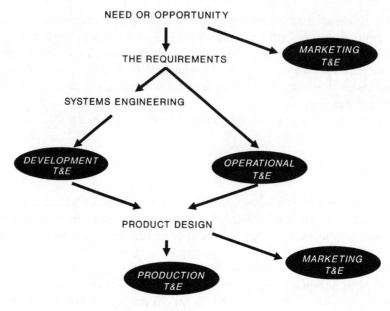

**Figure 4.3**
Flow-down of T & E requirements

some margin for making changes and to account for differences in test environments. For example, system reliability would be expected to be higher in a controlled demonstration in a somewhat benign factory environment than it would in its user environment. The overall system T & E criteria should have stabilized at the beginning of the engineering and manufacturing development phase to counter the well intended suggestions of people who, in an effort to clarify system performance requirements, actually recommend changes that would increase them. Such changes to the requirements after the engineering and manufacturing development phase has begun are disruptive, do a disservice to the disciplined systems engineering that has previously been accomplished and can invalidate the testing that has been conducted to date.

## 4.2   DETAILED PLANNING

### 4.2.1   The Scope, Timing, and Schedule

How much T & E is enough? Making this judgment is perhaps one of the greatest challenges for any development program. Figure 4.4 characterizes in a greatly exaggerated but pertinent fashion, the evolution of thought on this subject over the

‹1970   IT DOESN'T MATTER; NO ONE'S WATCHING ANYWAY.

1970'S   IF IT AIN'T BROKE, BREAK IT.

1980'S   IF YOU CAN'T TEST IT, SIMULATE IT.

1990'S   IF YOU DON'T BELIEVE THE SIMULATION, TEST IT.

›2000   DON'T BOTHER WITH T&E, THEY CAN ONLY AFFORD
        ONE OF 'EM ANYWAY.

        - - *BUT* - -

ALWAYS: HOW MUCH TESTING IS ENOUGH ?

**Figure 4.4**
How much T & E is enough?

past few decades (Greenlee 1992). In the '70s, the objective of most T & E programs was full-up testing of the entire system, in its planned environment. Because of complexity and cost of testing, in the '80s, modeling and simulation were used as tools to reduce the need for some of the more costly testing. But then the simulation sometimes proved untrustworthy, so in the 1990s, the models were to be fully validated with actual operational data before they could be used. There are no widely accepted methodologies for determining how much T & E is enough. By default, a program manager is sure that sufficient T & E is planned when all of the users of the T & E results are satisfied. Since T & E is both an information generator and risk reducer, T & E should stop when enough info has been gathered and the risk is at an acceptable level. Who defines 'acceptable'? The sponsor of the program, the system development team (the program manager and his technical agents, contractors, and subcontractors), the prospective users, consumer advocacy groups, and inspection agencies. In the case of government programs, the review/audit community, the funding appropriators (such as the US Congress), the taxpaying public, and the media also sometimes get to comment on what they consider 'acceptable'. So the primary drivers for the scope of T & E are the risks to be reduced and the satisfaction of the customers.

In the commercial arena, there are some well known cases of poor scoping. Scoping T & E of today's computer-intensive systems has become particularly difficult. In July of 1991, the East Coast of the United States suffered a massive phone outage, caused by three faulty lines of computer code – a little 'bug' in a software patch being incorporated into the telephone switching system. The testing that would have uncovered the 'bug' had not been conducted because of time and cost considerations. But the financial consequences of undetected deficiencies can overshadow the costs of T & E. American Airlines estimates that it cost $20,000 a minute for outages in their SABRE reservation system. A 12 hour outage in May 1989 cost over $14,000,000. Figure 4.5 lists costs of system down time of other software-intensive systems (Bender 1992).

---

### Case example: Chrysler mini-vans

In the 1980s, Chrysler Corporation, whose reputation had been made on large luxury cars, came close to bankruptcy because of dramatic shifts in customer automobile preferences. By the early 1990s, in various cost saving measures, Chrysler had cut back on funds available for design ... and testing. Its reputation for quality suffered, and that impacted sales. Chrysler then made a dramatic change in direction: it developed some of the most extensive T & E programs of any auto maker. Its mini-vans had a reputation for poorly fitting parts resulting in many squeaks and rattles. So Chrysler hired a company from Toronto, Canada, to shake prototype models of its mini-vans in a huge hydraulic device. The device simulated 99 400 mi. (160,000 km) of driving at temperatures ranging from 50 °C (122 °F) to minus 30 °C (minus 22 °F). Then Chrysler installed hydraulic shakers in its two plants to test vans coming off the production line – something no other car maker does. In addition, Chrysler constructed new Scientific Test Facilities next to its design facilities in Michigan, at an investment of over $140 million. These facilities include a 3/8-scale wind tunnel (Figure 4.6), an Environmental Test Center (Figure 4.7), and Electromagnetic Compatibility Center, Figure 4.8 (becoming more important as cars can use up to 20 microprocessors to control systems such as air bags, antilock brakes, and engine controls), Powertrain Test Facilities, and a Noise Vibration and Harshness (NVH) Laboratory (see also Chapter 6, Figure 6.15). Another improvement Chrysler made to its T & E program was the imposition of stricter testing requirements on its suppliers, after it learned that when those suppliers' production lines were fully operating, parts quality problems occurred. So Chrysler engineers now actually participate in test runs of parts assembly line operations at the vendors' plants to help spot problems early. As a result of these changes, Chrysler quickly made quality improvements surpassing most of the industry (Woodruff 1995).

---

- *American Airlines SABRE reservation system:*
*$20,000 per minute*

- *Boeing: $50,000 per minute*

- *American Express: $167,000 per minute*

- *Charles Schwab: $1,000,000 per minute*

**Figure 4.5**
Costs of system downtime

**Figure 4.6**
Chrysler 3/8-scale wind tunnel

**Figure 4.7**
Environmental Test Center (Climate)

**Figure 4.8**
Chrysler's electromagnetic compatibility testing

T & E events must complement each other. In complex development programs, it is very challenging to ensure that the overall T & E program is both cohesive and efficient. Without special efforts to cohesively integrate the T & E program, it will only be an aggregation of component, system, interface, and integration programs, optimized by those who have responsibilities for each subsystem or at each of those levels of design. Also, adding complexity to and risking inefficiency in the T & E program are the demands of so many diverse parties: the organization tasking and funding the T & E, quality assurance people, the hardware suppliers, the computer software developers, the program manager, the corporation, the customer, consumer groups, government inspection agencies, and, in the case of government-funded programs, government audit agencies and perhaps even Congress. One can see the potential for large T & E programs, with much unnecessary duplication and yet with notable deficiencies in coverage. A substantial effort needs to be undertaken to engineer testing so that each event complements the others, and the results are traceable between events.

There must be a very systematic approach to the test event sequencing. Even independently conducted full-scale operational T & E events must build upon the developmental T & E already conducted. Which tests – or which parts of each test – are prerequisites to which others must be known in advance, and well documented for all to be aware of. Likewise, *entry* and *exit* criteria for each test event should be identified, agreed to, and adhered to. This will allow the best scheduling work-arounds to be obvious when serious technical or resource problems are encountered.

Engineering the T & E events must accommodate not only the logical needs to minimize system performance risks, but also the overall *T & E program structure*. By the 'structure' we mean how the major T & E events are scheduled and conducted to support the points in time when the significant design or investment decisions are to be made. The objective is to optimize the structure so that the technical and financial risks to be taken at those decision points are minimized. Program and progress reviews should be judged against pre-established criteria, including T & E requirements. The milestone procurement process used by some US government departments and similar systems of checkpoints used in some commercial developments (e.g., 'tollgates' used by General Electric in its aircraft engine programs) are examples of how T & E can be used to directly pace a development program. T & E events should be tailored in scope and timing to achieve this objective.

The actual scheduling of individual tests should involve a building-block approach so as to progressively build confidence in the system. Concentrating the focus of major test events at the heart or at the extremes of the performance envelope should be avoided. The results would skew the conclusions that are drawn, particularly by outsiders who have time only to review the top-level results.

While the T & E program is one pacing item for the introduction of new products, it is not the only one. There are pressures to field or sell systems as soon as possible. For military systems, the need to be able to counter the ever-increasing threat is always a race against time. For commercial systems, there is the pressure to be the first to get to the market in order to capture as large a share as possible and thereby provide to the investors a good return on their investments. For all programs, there is a need to meet schedules and stay within budget constraints. So planning a T & E program is not just structuring tests to verify that the system meets its requirements. The scheduling and scoping of the T & E become key drivers of whether or not the T & E program *per se* will be successful in its objectives.

## 4.2.2   Engineering the T & E Events

The five key steps in designing a test event are shown in Figure 4.9. A test event usually spans a period of time (from several hours to several weeks) during which a set of test procedures will be conducted on a particular configuration of the system in order to accomplish specific objectives, such as to verify proper performance, to measure strengths and limitations in the system's capabilities, or to validate the predictions of a model.

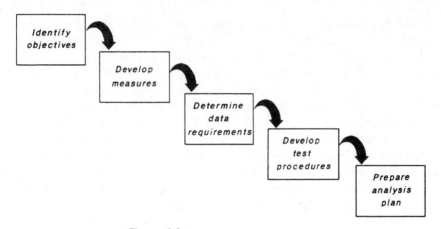

**Figure 4.9**
Steps in the design of a test event

First, the *objectives* of the test event need to be identified as specifically as possible. By the time detailed planning for the event commences, the objectives will probably be well defined through top-level planning in a T & E Master Plan (see Chapter 5). To whatever extent they need to be further refined, this must be done before any other planning proceeds, so that the time and resources for each part of the event will be judiciously allocated in a well balanced manner. The objectives should also be prioritized, so that if resources, scheduling or technical problems occur, the optimum work-arounds that lessen the risks and the impacts on the program schedule can be arranged.

Secondly, the full list of *parameters* to be measured must be determined. The top level measures – the predominant ones that characterize the key performance features of the system – should readily flow from the system requirements documents. All other parameters to be measured should then be identified and tiered to show their interdependencies. A parameter tree or a matrix may provide a useful format for displaying these parameters. Both quantitative and qualitative measures should be identified in this manner. For each parameter, evaluation criteria should be identified. The criteria that are quantitative should be well identified to the extent necessary for analysis and evaluation. For example, a single threshold number (pass/fail), or a band of values which bound the acceptable results, or a table of values that are acceptable under different combinations of conditions may be appropriate. These criteria should be traceable to the system performance requirements over the full range of operational conditions, including manufacturing, test, transportation, and storage, as they apply to the system performance profile. The system modeling and simulations used in design should be used in conjunction with stress analyses, failure modes and effects analyses, worst case analyses, etc., to help identify the more important parameters and the related criteria for use in the various test events in the program.

T & E programs for complex systems frequently necessitate the development of new measurement systems. The two disciplines of measurement science and instrument engineering have as their shared objective the adequate measurement of everything necessary. The careful observation of the 'signal', the calibration of instrumentation, the collection of the data, and the transformation of the data into information and knowledge are critical to a successful measurement system. The reader is referred to other books in this series, in particular 'Introduction to Measurement Science and Engineering' (Sydenham, Hancock and Thorn 1989) for an excellent tutorial treatment of both the formal and heuristic concepts that underlie this specialized field.

Thirdly, the *data requirements* need to be finalized. Given the cost of testing, it is vital to judiciously determine well in advance of the test how much data needs to be taken. A typical approach in the past has been to take as much data as possible, from which the more important data can be selected later during the evaluation effort. That is no longer an affordable approach for today's complex systems. For example, in the communications industry, there is now a standard way of transmitting voice, video, and data in 53-byte cells that provides speed, flexibility, and compatibility with many different applications. The high bandwidth and unusual flexibility of this 'asynchronous transfer mode' (ATM) have created new challenges for T & E engineering. Hardware-based cell-switching fabrics small enough to fit into a personal-computer sized box have capacities measured in gigabytes per second. To put this in perspective, realize that 1 gigabyte per second of switching capacity is sufficient to handle more than 15,000 voice connections simultaneously (Ruiu 1994). Data needs will drive the cost of special instrumentation, computer analysis time, test time, and evaluation time. They must be well engineered. The selection of test methodologies and the derivation of data requirements are inseparable and of equal importance. Well executed tests with inadequate data collection are as useless as irrelevant tests with excellent data collection.

Fourth, the *test methodologies* need to be developed, and the detailed step-by-step *test procedures* need to be documented, validated and published. They should be stand-alone to the maximum extent possible, versus dependent on many other references for a true understanding of the requirements. Being stand-alone will enable and even encourage reviews by many parties. Entry criteria to start the test, suspension criteria to be used during the test, and the accept/reject criteria should be documented in the test procedures. In developing test methodologies and test procedures, conscious decisions should be made to determine the best mix of inspections, mathematical extrapolations, demonstrations without substantial data collection, and actual system operation with well planned measurements and data collection. In production acceptance T & E in particular, such as lot acceptance testing, the best mix may change over time, so that the T & E of early production units may rely more on end-to-end demonstrations of system functions and the T & E of later units to a larger extent on less costly inspections. The criteria for selecting test methodologies should not be based only on the cost of development or the cost of conducting them. To the greatest extent possible, test methodologies and test procedures should be validated through prior use. Use of T & E standards published by government agencies and professional and trade associations (see section 4.3.1) is one way of having confidence in the validity of the methodology. But care must be taken to ensure that differences in the conditions under which the test will be conducted compared to the conditions identified in the standard, or even obsolescence of the standard, do not undermine that validity. One might think for example that automobile crash testing involves standard procedures that rarely change. This is not the case. Aside from what is being learned from the increasingly sophisticated instrumentation and crash dummies, as well as continuing world-wide research, the increasing use of passenger restraint systems and changes in the structural design of automobiles make this a rapidly evolving field. A few decades ago, tests usually involved crashing the full fronts of vehicles into full-width rigid barriers at 30 and 35 mph. But it was learned from research led by organizations such as the Insurance Institute for Highway Safety (see Section 1.3) that full-width crashes rarely involve intrusion of the occupant compartment, which does occur in offset crashes. Tests are now conducted with the automobile's front end hitting the edge instead of the center of a barrier, as well as with barriers with deformable faces. The latter tests make the car responsible for directing the impact forces into the correct parts of the structure as would happen in automobile-to-automobile crashes, removing the unrealistic deceleration effects from rigid wall tests.

Fifth, and most importantly, an *analysis plan* should be prepared. How much and what types of data will be available, how it will be analyzed, and what analysis tools will be needed all need to be planned in advance. The methodologies for analysis must be well thought out and verified. The entire T & E team, and perhaps the requirements setter and the user, should be aware of and agree with these methodologies. To the greatest extent practical, these methodologies should be validated in advance. The process of developing the analysis plan will probably prompt changes to the data collection plan, to the parameters to be measured, and perhaps even to the objectives of the test event. The development of the plan may even suggest an addition to the objectives of an earlier test event, for example, validation of a data analysis methodology.

Section 5.3, discusses the format for a test event plan that describes the items that need to be documented to capture this information and maintain it current.

## 4.2.3   Statistical Design

T & E of complex systems cannot be effectively or efficiently planned without the use of statistical design and analysis. The statistical design of a test can impact the objectives of the test program, the resources needed to support the tests, the set-up conditions for each test, and the data analysis plan. The first step in statistical design is to identify the dependent variable characteristics. Essentially, this is what will be measured in the test. Secondly, the independent variables are identified, as well as the range of values that they should be subjected to during testing. Then the type of statistical problem is identified (e.g., estimations, comparisons or determining relationships between variables) for which candidate solution techniques can be selected. The major challenge is to select the techniques that get the most information related to the test objectives out of the data that can be affordably collected and analyzed. If the collection and analysis of data has a very firm constraint (e.g., a limit on the number of test firings of satellite launch rocket motors), a variety of techniques can be used to optimize the statistical designs and corresponding sample and replication sizes. The dependent variables will be somewhat obvious to the systems engineering team; they will equate to, or actually be, the parameters that are identified in the system requirements and the Test and Evaluation Master Plan. However, judiciously identifying the independent variables is more challenging. Financial and scheduling resources can be significantly saved or misspent as a result of decisions to control a particular independent variable during a test, and – if controlled – to decide whether or not it is to be treated as a primary factor, or as a background factor that is to be held fixed or to be ignored (but nonetheless measured), see Figure 4.10. Neither the effectiveness nor the efficiency of a test event can be assured if statistical design is not a part of the planning.

## 4.2.4  A Focus on the Operational

The T & E program should be engineered into a building-block approach from detailed, controlled, developmental T & E to more 'free play' full-scale

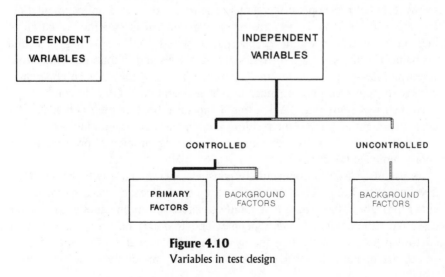

**Figure 4.10**
Variables in test design

operational T & E. The build-up from developmental to operational T & E should not be discrete, but should be a smooth transition. To the maximum extent possible, developmental T & E events should help achieve operational T & E objectives, and operational T & E events should add to the database of information on system capabilities and limitations captured by developmental T & E.

To achieve the objectives of the operational T & E, it must be conducted in as realistic a setting as possible. The primary value of operational T & E lies in the simplicity of the concept. According to the US Navy's Operational Test and Evaluation Force, operational T & E of Navy systems 'is conducted to prove a system's capabilities in a realistic operational environment, when operated and maintained by sailors, subjected to routine wear-and-tear, and employed in typical combat conditions against an enemy who strikes back' (US Navy 1992). In contrast to developmental T & E, which deals mostly with instrumented tests and statistically validated data, Navy operational T & E deals primarily with the operational realism and the uncertainties of combat. 'The objective is not always to acquire statistically significant data, or a box score of successes or failures (since replications are seldom possible), but to gain the most complete understanding possible of the system's capabilities under stress ... . In operational T & E, the principal value derived is often unplanned, resulting not from the basic purpose of the test, but from realistic aspects ... that were not expected to occur.'

It is important to ensure that operational T & E is conducted as early as practical and throughout the program. The operational T & E events that are conducted in support of production decisions in US Defense Department programs are run as dedicated exercises, independent of the planning, testing, controls, and even witnessing of the government and contractor development teams. Rather than being a real 'test' in the classical engineering sense, such events are operational exercises in which not only the hardware and computer software, but also the operator and maintenance people, the user documentation, training, and logistics support are being evaluated. It is imperative that all of these items be fully ready for the test so that the test conductors can give a valid assessment of the system's operational capabilities and limitations. Otherwise, when problems do occur during operational T & E, it may not be apparent – because of the 'free play' manner in which operational T & E is conducted – what their real sources are. What might appear to be a training deficiency, for example, might instead be the result of incomplete troubleshooting procedures, immature computer software, human engineering problems, or even a few poorly motivated operators.

As useful as it is to conceptually consider operational T & E distinct from developmental T & E, it is a mistake to treat the actual test events associated with either throughout the program as totally separated from one another. In such a case, the developmental T & E people tend to focus only on running very controlled tests to verify that the system and components meet specification values, and operational T & E people tend to focus only on free play exercises in which they observe the prospective user operating and maintaining the system 'as he would if he owned it.' The most obvious problem with this approach is the increased risk that when the system is turned over for operational T & E with user/customer involvement, it will fail. Of concern also is the risk that operational limitations will not be detected before the system is deployed. Because it occurs late in development, operational T & E is conducted under much greater time and scheduling constraints than developmental T & E. Considering the multitude of environments, conditions and scenarios the system may see in actual usage, operational tests can cover but a small snapshot. Operational T & E should be structured to build upon developmental tests, some of which in themselves are operationally oriented.

## 4.3 RESOURCES

Having the proper resources in place at the proper time is critical to a successful T & E program. Such resources include T & E standards, test facilities (instrumentation, test articles, expendables, and test sites) and people.

### 4.3.1 T & E Standards

There is an ever increasing number of T & E standards and specifications that are readily accessible. Appendix A lists the major world-wide organizations that publish and maintain T & E standards, as well as the addresses for ordering copies of their standards. In addition, there are many standards maintained by government agencies and commercial organizations to which access is restricted because they contain sensitive information of military or commercially competitive value. Such documents are not as easily identified, but T & E managers with a 'need to know' can, with some research, determine where and how to obtain access to them. While using standard test methodologies can save the cost of developing new tests and can increase confidence in the validity of the T & E itself, care must be taken to ensure that the standards can be correctly applied and are current. See section 4.2.2.

### 4.3.2 Test Facilities

Full-scale system tests frequently require unique resources. For example, there is often a need for special test stands and instrumentation; and arrangements for data analyses, operational and maintenance personnel; and training for those operating the system during T & E as well as for the members of the T & E team themselves.

For military systems, munitions, specially configured targets and access to national test and training ranges may be necessary. To arrange to have all these available at the appointed time requires careful planning and diligence. Some, such as new instrumentation, may necessitate major development efforts themselves, that must begin concurrently with the system design if they are to be available for the full-scale developmental and operational T & E. Others, such as the Defense Department's T & E and training ranges and federal government laboratories, are national assets that must be scheduled well in advance of the T & E. The ranges owned by the US Defense Department represent a capital investment of well over $20 billion, and encompass 21,000 square miles of land, 243 square miles of sea space, and 221,000 square miles of air space.

---

### Case example: Ohmsett oil spill response test facility

In 1974, the Ohmsett National Oil Spill Response Test Facility opened in Leonardo, New Jersey. Built by the US Environmental Protection Agency, it is now managed by the Department of the Interior. Today it remains a one-of-a-kind facility where full-scale oil spill response equipment can be conducted with oil under controlled conditions including varying wave simulations. The facility features an open-air tow and wave tank (Figure 4.11) where environmentally safe testing can be done in the presence of oil. The tank is 203 metres (667 feet) long by 20 metres (65 feet) wide and 2.4 metres (8 feet) deep. It has a towing carriage system with four bridges to pull devices under test, a wave generator capable of simulating either regular or confused waves, a partially moveable beach at one end to control wave reflection, and chlorination and filtration systems. Ohmsett's size allows various oil spill recovery components (booms, skimmers and temporary storage devices) to be combined into systems that can be realistically evaluated and compared with each other. Figure 4.12 shows T & E of an oil skimming system being sponsored by the US Coast Guard and the Canadian Coast Guard.

---

Today, the actual site of T & E facilities can in some cases be almost incidental to the location of the design sites. With the capabilities available today for accurate and real-time data transmission, the development people do not have to be co-located with the test site. In the early 1980s, McDonnell Douglas had owned large cryogenics vessels, but had not relocated them when it moved from its Santa Monica plant to Huntington Beach, California. Several years later, when they decided to move the vessels, they could not do so because of the environmental risks of locating such facilities near a heavily populated urban center. They contracted for Wyle Laboratories, Inc., to take over the facilities, and to design, build, set up and operate them 40 miles away from Huntington Beach. Technology now allows us to transmit over 2000 channels of data via a dedicated satellite transmission system for less than $2000 a month.

**Figure 4.11**
The concrete tank at Ohmsett

**Figure 4.12**
T & E of an oil skimming system

With proper planning, a test can be remotely accomplished and the design staff can have access to the data directly or via a link, paperless and in real time.

As systems have become more complex, the performance of the subsystems has become more interdependent, and higher degrees of integration have been achieved. System integration test sites have become useful and worthwhile investments for identifying integration engineering problems earlier and thereby saving time and money, and perhaps the risk of a suboptimized design, when production commences. For development programs for military systems, such sites have been used since the 1970s as intermediate staging locations in which to verify system integration and debug the integration of computer software, before systems are installed during the construction of the first host ships and aircraft. This is particularly true for combat systems, where a number of different sensors, fire control systems, and weapons are integrated through a single command and control system. Such sites have proven to greatly reduce the time and cost of testing during the ship or aircraft construction periods. Even after the first ships and aircraft have been delivered, these sites are frequently kept open to test each set of systems on their way to follow-on platforms, and for resolving emergent problems that occur in systems that are in service. The Federal Aviation Administration also has a major integration test site at its Technical Center in Atlantic City, New Jersey. System additions and changes to the systems at the air traffic control centers in the United States are thoroughly

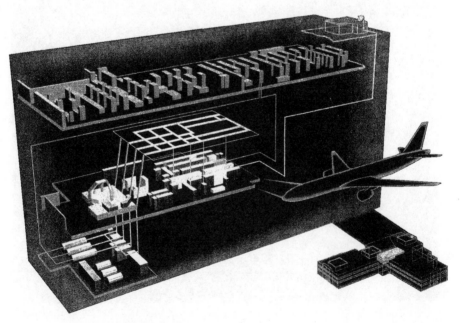

**Figure 4.13**
Boeing's 777 Integrated Aircraft Systems Lab

tested – both in development and operational testing – before they are fielded. As part of Boeing's development program for its new '777' aircraft, Boeing built an Integrated Aircraft Systems Laboratory at its plant in Seattle, Washington to conduct T & E on avionics, hydraulics, electrical systems, pneumatics, fiber-optics and flight control electromechanical interfaces. The upper level houses the simulation and test equipment; the remainder of the equipment below is actual aircraft systems of the same configuration as installed in the prototype aircraft and eventually the production aircraft. This facility helped minimize expensive production floor change and cut warranty costs. More about the Boeing 777 T & E program is described in Section 7.1.

Another support resource the availability of which is critical to T & E is modeling and simulation. This work needs to be planned well in advance so it can have some credible level of validation prior to use in direct support of T & E. The operative word here is 'credible'. Section 6.11.4 discusses the validation of models and simulation techniques.

## 4.3.3 The 'People' Resources

Those planning and conducting the T & E program need to be effective managers, good systems engineers, and have training and experience in T & E. With the advent of T & E professional associations like the International Test and Evaluation Association, ITEA, (see Section 8.5), information on the availability of such people is becoming easier to obtain. Appendix B lists the major qualification requirements for T & E personnel.

Aside from the T & E people themselves, there are many important participant organizations which have vested interests in the T & E program: the program office, the prime contractor, subcontractors, suppliers of equipment (vendors), design organizations, test laboratories, the customer, and the financial sponsor. They should all be involved in its planning, or at least agree to it before it is executed. The formal contracts between the parties must not hinder but must induce close, open, and 'real-time' coordination among those involved in the T & E program. The T & E people in each organization need to be networked to be able to share plans, information, problem analyses, and test reports. An overall T & E management plan for the program should be developed to describe the roles and responsibilities of the major participants, points of contact, reporting requirements, and schedules. This differs from the T & E Master Plan, which is an executive-level contract between the major parties in the program. The development of this plan can be best led by the program office, but it should be coordinated with all participants. The T & E program team should meet regularly and use the management plan as a point of departure in discussing progress, problems and future plans. The communication of T & E information must maintain high standards for both accuracy and timeliness. Successes and problems must be thoroughly identified, analyzed, and compared to past experience (i.e., trend analyses). All participants need to be kept aware of the progress and problems, and have easy access to any detailed information they might need. In any complex T & E program, there will always be a certain amount of subjective judgement necessary. It is important that there be an openness among the T & E participants to instill mutual respect and overall confidence in those judgments. In short, those involved in the T & E must think of themselves not only as part of their own organizations, but also as a part of a T & E TEAM, networked to take maximum advantage of the very valuable information provided from the results of T & E (see Figure 4.14).

As part of the 'teaming' of the participants in a T & E program, it is frequently useful to have some amount of organizational separation for selected T & E people from those responsible for the design and development work. As discussed in section 3.2.3, the value of independence of at least some T & E people in each program is becoming recognized more and more. If the T & E focused only on verifying compliance with specifications, such independence would not be necessary. But as described in this book, T & E today requires an understanding of both how the system will likely be used and of the customer's needs. To have a T & E program controlled solely by the designers will likely result in many of the designers' biases being mistakenly overlooked, but to have it controlled solely by independent testers will likely result in it being inefficient and unnecessarily expensive. The challenge in the T & E program is to balance

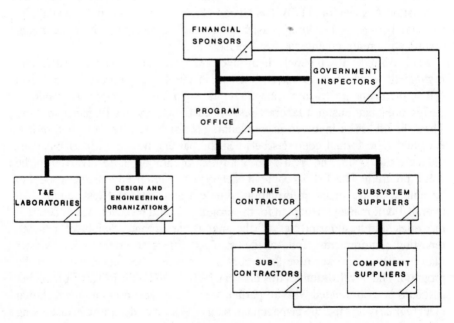

**Figure 4.14**
Networking the T & E people

the knowledge of the designers with the objectivity of some independent T & E people. Although there will be some independence, everyone working on the program, regardless of their organizational placement and mission, still needs to work in an integrated fashion to bring the best possible system into service or the best possible product to the market-place. They should work as a team, taking pride in the credibility of the T & E program and its contribution to the overall development program. The team should be respected for both its professionalism and objectivity by the rest of those involved in the system's development.

## 4.4 TEST CONDUCT

The basic rules for conducting T & E are so simple, yet it is easy to lose site of them in a complex system T & E program. Theresa and Patrick Reynolds, considered by some to be child prodigies, were conducting science experiments intended for ten-year olds when they were a mere three and four years old. The rules they learned from the primer they used (Walpole 1995) for conducting experiments also constitute 'Reynolds' Seven Rules' for conducting a T & E event:

ONE – Collect together all the things you need. Ask permission to use things that are not yours.

TWO – Keep your equipment separate from your family's.

THREE – As you try the experiments, write down or sketch exactly what you do and see .... If you want to try again, it is useful to know what happened the first time.

FOUR – Watch all ... very carefully.

FIVE – Always clean up after you have finished.

SIX – Don't worry if your answers are not exactly the same as those in the book. Try to work out what happened and why.

SEVEN – Be careful! Some experiments can be dangerous.

The T & E should be conducted to the greatest extent possible according to the agreed-upon plans. When there is a need to deviate from those plans, all parties that it could impact should be told – in advance if possible. For selected types of deviation, review and approval authorities should be identified. Deviations from the planned system configuration and test set-up in particular should be analyzed for impact on the validity of the T & E. Test schedule, test scenarios, and test procedures should be kept updated. The test entry criteria and test suspension criteria should be adhered to. Test problems should be recorded and dealt with expeditiously. Changes in the data recording or analysis should be well documented and reviewed. Although discipline and rigor must be maintained during the testing, the T & E conductors nevertheless should be allowed the time and flexibility to 'freelance' as they deem useful, as long as it does not interfere with the preplanned portion of the testing. All such excursions and the results should be carefully documented.

## 4.5   THE FEEDBACK LOOP

A T & E program needs to have a good closed-loop failure reporting, analysis, and corrective action system. That system will need to be modified to suit the development phase, the production phase, and the in-service phase:

• For the *development* phase, the system needs to be closely linked to the various sites where T & E of subsystems and full-scale production are being conducted. The primary objective of the system during the development phase is to carefully document the problems, the conditions under which they were evident, and the full effects of the fixes – so that if they reoccur at a later time or in a different location, the failure analysis can benefit from the previous work. As a development program proceeds, the identification and resolution of problems become significantly more costly, so anything that reduces the cost (and time) of later problem resolutions has a high pay-off. Besides, it is in the early part of the development that the program has the most flexibility in incorporating fixes, and therefore can readily optimize and harmonize them to the system design. The failure reporting system used during development will need to allow for complete descriptions of the problems or symptoms; discussions of root cause analyses, and detailed discussion of the effects of fixes – both those that appear ineffective and those that work. Documenting such detail involves a significant expenditure of work hours, and doing so accurately should include a good scrub by several participants that have seen the problem first-hand. Once the problem and resolution are reported, they need to be integrated into the program database to allow for comparison and possible correlation with other problems that might show evidence of systemic problems not otherwise uncovered.

• For the *production* phase, the numbers of sites and system configurations are much smaller than during development, so the documenting of problems and solutions can involve less detail. Production work tends to be manpower-intensive and schedule-intensive, so the primary objective of the failure reporting system is to expedite problem resolutions. During production, the tracking of problems can provide meaningful indicators of technical progress. Assuming that quality assurance efforts remain the same, and the work force is stable, there should be a dramatic reduction, from the early products to the later ones, of the numbers of problems and the average work required to fix each problem. If production starts up in a different location, or if significant modifications are made to the system, the numbers can be expected to rise temporarily.

● After the initial systems are placed *in-service*, there should be some form of feedback of the numbers and severity of problems encountered, so that the design of the later systems on the production line – and the design of the next generation system – can benefit from them. If a warranty program is in place, it will provide some good feedback. Or if the producer is involved in supporting the systems as they are serviced, he can help capture and analyze in-service data. However, it is important that whoever collects the data makes it available to all appropriate members of the development team, including subcontractors.

In preparation for both the interim and final analyses of test results as well as for progress reporting purposes, it is useful to have in place a categorization of how the various types of T & E problems will be sorted. One possible categorization scheme is described in Section 5.4.

The three primary pitfalls that frequently make problem reporting systems ineffective are lack of fully integrated feedback, incomplete procedures at the working level, and insufficient discipline among the participants. An atmosphere of openness and cooperation needs to be encouraged in each part of the program team so that all problems are surfaced at the earliest time possible.

The root cause analyses themselves should be conducted by the design team with the assistance of the T & E team. Detailed records of such analyses should be kept for future reference if needed. Any recommendations for re-tests should be tracked to ensure adequate consideration is given to the recommendations.

Part of the post-test analyses should be a detailed review of all reported problems to determine how to improve the T & E process in the future.

## 4.6 ANALYSES AND REPORTING

If the T & E event has been planned thoroughly (as described in sections 4.2.3 and 4.2.4), the data analyses *per se* can be done according to well defined and validated procedures that focus on the deviations and variances of the test results from what was expected. But this is only one of three factors in the evaluation part of T & E. The other two are assessments of the find-and-fix rate for problems during the T & E event, and the evaluation of the seriousness of the problems. While benchmarks for all three of these factors should be prepared in advance, the latter two will by their nature be primarily subjective. Nevertheless it is the analysis of all three factors together that enables meaningful evaluations to be made about how well the system has performed and the validity of the T & E itself.

Drawing conclusions from the analyzed test data must be done logically and

**Figure 4.15**
Derivation of conclusions

carefully. The conclusions should be able to be substantiated from the data either by deductive or inductive reasoning, and it should be clear whether each is based on

- the test results alone

- the test results and the results of other evaluations

- the test results and knowledge of the practices and usage in the users' environment

- the test results and the knowledge the system requirements, or

- some combination of the above.

If the test was properly conducted, it would seem that conclusions could be drawn almost mechanically. However, in practice, there is a more involved process of drawing inferences from test results. Conclusions are actually drawn in four steps as depicted in Figure 4.15 (USNSWC 1985).

The formulation of conclusions and recommendations typically requires a considerable amount of time. To fulfill their purposes, they need to meet the needs of the primary audiences for which they are written. The timing and content of the reports should be identified well in advance of the conduct of T & E, and tailored to the expected readership. The useful format for the report of a T & E program event is described in Section 5.5.


## 4.7  SUMMARY

It is impossible to meaningfully consider a 'cook book' approach to T & E engineering. For it to be both effective and affordable, T & E must be tailored to both the type of system and to the risks (technical, scheduling, and cost risks) inherent in the particular program. Disciplined engineering coupled with sound technical judgment is the key to effective planning and execution of the T & E program.

# REFERENCES

Bender, R. (1992) of Bender & Associates, at the June 1992 workshop on computer software testing co-sponsored in Washington, D.C. by the American Society of Quality Control and the International T & E Association.

Greenlee, D. R. (1992) Vice President, Science Applications International Corporation (SAIC) and Vice President of the International Test and Evaluation Association (ITEA), *presentation at the October 1992 Annual ITEA Symposium in Albuquerque, New Mexico*, ITEA, Fairfax, Va.

Kass, R. A. (1995) Writing Measures of Performance to Get the Right Data. *The ITEA Journal of Test and Evaluation*, **16**, (2), pp. 12–16.

Ruiu, D. (1994) Testing ATM Systems. Institute of Electrical and Electronics Engineering *Spectrum*, June, p. 25.

Sydenham, P.H., Hancock, N.H., and Thorn, R. (1989) *Introduction to Measurement Science and Engineering*. Wiley, Chichester.

US Navy (1992) *The Operational Test Directors' Guide* US Navy Operational Test and Evaluation Force Instruction 3960.1G, January 2.

USNSWC (1985) *Advanced Surface Ship Weapons Systems Test and Evaluation Guide*, Revision 1, US Naval Surface Warfare Center Port Hueneme Division, October.

Walpole, B. (1995) *Science Fun Station* Price Stern Sloan, New York.

Woodruff, D. (1995) An Embarrassment of Glitches Galvanizes Chrysler. *Business Week*, April 17.

# 5

# T & E Program Documentation

*The frightening thought that what you draw may become a building
makes for reasoned lines*

Saul Steinberg

The need to properly capture and document what is planned, what actually happens, and the interpretation of both has become critically important in every aspect of the T & E of complex systems. The ability to collect, process, store and analyze huge amounts of information both accurately and relatively inexpensively has enabled technology to be applied so rapidly in the last few decades. To handle these processes systematically in T & E programs requires a well controlled and widely understood hierarchy of documentation. This chapter discusses the top-level documents needed for the proper T & E engineering described in Chapter 4.

## 5.1 THE T & E MASTER PLAN: 'THE CONTRACT'

The need for a 'master plan' for a T & E program has evolved from the increasing number of participants with varying but nonetheless direct interests in the program. For example, the user wants assurance that the T & E program is realistic and therefore can be relied upon to have identified major technical and operational problems before a system is placed in service. The financial sponsors want the T & E to provide assurance that the user will be satisfied with the system, thereby obtaining a return for their investments. In addition, they want the T & E to the maximum extent possible to pace the investment of funding so as to keep financial risks at any time to a minimum. The companies that develop and produce the system want to make sure that the T & E identifies major problems early before they cause much rework and cut into profits, and that the T & E results are an asset for marketing the system, not a liability. The T & E team wants a T & E program that will be a solid basis for the conclusions and recommendations they are tasked to make. And the program office has to integrate and balance all of these sometimes conflicting needs. The TEMP is an executive level document that defines the T & E program in enough depth to assure the customers of the T & E program that their needs have been met. The TEMP is most effective when it is treated as a contract between all of these parties, and signed by each of them.

In the early 1980s, the Defense Department instituted the TEMP as the top level T & E planning document to be used in its acquisition programs. By the end of that decade, there were over a thousand active programs that had TEMPs. Even more than balancing the interests of the major participants listed above, the Defense Department's TEMPs' primary use was to effect and describe the integration of operational T & E into, and the involvement of independent T & E agencies in, each program (USDoD 1991). TEMPs for the highest visibility programs (about the top 5%) are approved by the Defense Department's Director of Operational Test and Evaluation (DOT & E), whose office was established as a result of Public Law that required such oversight to ensure thorough and independent operational T & E are conducted in each program. In addition, the Military Services have adopted the use of the TEMP for lower-level programs. The two T & E offices on the staff of Secretary of Defense, that of the Director of Operational T & E and the office of Director of Defense Test, Systems Engineering, and Evaluation, use the TEMP to review each program's implementation of three T & E policies in particular. First, formal, agreed-upon T & E is to begin early and be conducted throughout the acquisition program. System performance requirements, test objectives, and evaluation criteria are to be established and agreed upon before the testing begins. Secondly, successful accomplishment of the identified T & E objectives is a prerequisite for committing additional resources to a program and advancing it from one phase to another. The structure of the acquisition program is to be based on this principle. That is, the major decision points in each program, at which it is authorized to proceed to the next phase, are to be planned at points in the program where there are meaningful T & E results available. Thirdly, full-scale operational T & E is to be successfully conducted on a production-representative system before the program proceeds to full production. In addition, the Director of Operational T & E uses the TEMP to review implementation and adequacy of 'live-fire' testing in programs to which it applies (Section 6.9).

The TEMP should describe the overall T & E strategy in the program. If the program involves substantial financial risk, the TEMP should display the T & E requirements that must be met in relationship to the major investment decision milestones in that program. In other words, it demonstrates 'try before buy.' If the program involves a build-up to a single major event, such as the launch, operation, and recovery of a space shuttle, the TEMP should demonstrate that the T & E involves progressive building of confidence from earlier to later T & E events. If the program's objective is to field a new drug, the TEMP should show how the T & E program will prove that the benefits outweigh the risks of the drug. If the program consists in introducing in stages improvements to systems that are already in service and that must continue to be on-line, the TEMP should describe how the T & E will reduce the risk of interruption or temporary loss of capabilities. An example is the US Federal Aviation Administration's updating of the air traffic control systems.

Aside from describing the overall program structure, there are several other purposes, shown in Figure 5.1, that can be served by a TEMP. First, it can provide a description of the system and its usage profile in terms such that judgments can be made about the adequacy of evaluation criteria to be used in deciding its readiness for production and placement in service. Secondly, the TEMP can describe the scope of each of the major developmental T & E and operational T & E events as evidence that sufficient testing is planned to give decision makers confidence in system performance and the efficacy of the resources to be spent. It also describes for each event any limitations that will impact the T & E team's ability to draw conclusions or to make recommendations. The Defense Department places great importance on an accurate list of such limitations in the TEMP, since less than full disclosure can reduce the level of confidence in the results of testing. Both limitations in the test environment compared with the full range of environments the system will experience in service, and how representative the user personnel are compared to the total population of prospective users need to be known and agreed to early in the program by all organizations which will later be reviewing the T & E results in order to make design and programmatic decisions. Third, the TEMP can describe aspects of the program that must receive special attention and approvals, such as safety certifications. Fourth, it should describe the planned use of modeling and simulations in support of the T & E program, and how they will be validated prior to such use. Fifth, it should be used to describe any special resources that will be necessary to support the T & E program, such as new instrumentation or access to special laboratory test chambers. These descriptions are necessary so that the resources are identified and agreed to sufficiently in advance, so that their capabilities and limitations are known, and so that the test conductors can properly plan for their usage. They are also necessary so that the financial sponsors and the program manager can budget for any such resources that he must acquire. In addition, the numbers and types of such resources give

- Serves as a contract between major participants.

- Displays phasing of T&E prior to major milestones.

- Describes the system & its performance requirements,
    in testable terms.

- Documents the scope of each major T&E event,
    including anticipated limitations.

- Describes planned use of modeling and simulation.

- Lists special resources required.

**Figure 5.1**
Purposes of a TEMP

insight to the program reviewers and decision makers into how complete the testing will be at various points in the program.

Ford Motor Company's Global Test Operations have used a Design Verification Plan and Report as their T & E Master Plan in some of their automobile development programs.

Figure 5.2 illustrates how the TEMP interrelates with other key program documents. In particular, the system performance requirements in the TEMP evolve from and expand upon the System Requirements Document. As system engineering proceeds, a series of specifications is developed by the development team. The specifications, and in particular the performance requirements and test requirements, should be traceable to the TEMP. The specifications are invoked in the tasking orders within the company and in contracts with others. From these specifications, the system integration designer and the subsystem developers prepare their T & E plans. Also, the program manager, the laboratories and the engineering activities develop their T & E plans from the specifications. In order to ensure that full system performance is demonstrated, the T & E may sometimes exceed the specification requirements to test the system on the fringes of or outside its performance envelope to ensure that all T & E risks are addressed, and to learn more about the capabilities and limitations of the system.

## 5.2 CONTENTS OF THE T & E MASTER PLAN

The TEMP should be an executive-level document that describes the T & E program objectives and shows the planned mix of the various types of T & E and T & E events that will be used to accomplish those objectives. It should contain:

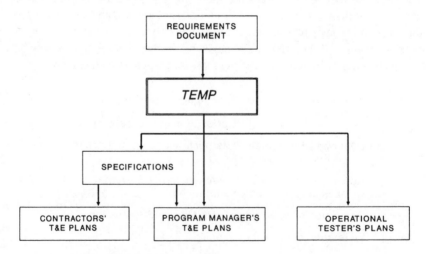

**Figure 5.2**
Relationship between key T & E program documents

(A)   a description of the system, its unique capabilities and how they will differ
       from what is already in service, the high risk aspects of the design, the key
       program thresholds which, if not achieved, might cause the program to be
       cancelled, and other metrics that the T & E conductors are to measure and
       report on.

(B)   an integrated schedule showing the major T & E events in relation to other
       program events, such as financial commitments, progress approvals,
       contract awards, and system procurement and delivery schedules.

(C)   a narrative description of each major T & E event (objectives, scope,
       methodology, resources required, system configuration, modeling and
       simulation to be used, and limitations to the scope of T & E). These
       T & E events can be described in the sequence in which they will occur, or
       be grouped by locations or by type of test (e.g. developmental and
       operational), or be separated by test location or test conductor
       (development company facilities versus outside specialized test facilities).

(D)   a list of the special resource requirements to support T & E events.

The US Department of Defense dedicates the first part of a TEMP to a
description of the system and the performance requirements, the second to an
integrated schedule, the third to developmental T & E, the fourth to operational
T & E, and the fifth to special resources.

   A well engineered design will have different levels of performance requirements (system, subsystem, component, etc.) whose linkage to each other will
vary. Some will be threshold values that reflect a dominant characteristic of the
system that is a design driver (e.g., the reliability of an air traffic control system
or the range of a military weapon system). Others will be second or lower tier
numbers derived from the top-level requirements but perhaps not absolutely
necessary for the achievement of those requirements. Capturing the interrelationships of these parameters and assigned values as the system design evolves
and design trade-offs are made is a difficult but necessary task. Today, there are
good computer-based metric correlation tools available that are well suited to
this type of analysis.

   Some parameters of interest may not be easily quantifiable. Of course, they
will be easier to use in the T & E program if they are quantified. Lord Kelvin
once observed, 'When you can measure what you are speaking about, and
express it in numbers, you know something about it; but when you cannot
express it in numbers, your knowledge is of a meager and unsatisfactory kind. It
may be the beginning of knowledge, but you have scarcely, in your thoughts,
advanced to the stage of science.' However, some important characteristics such
as logistics supportability and human factors engineering can only be quantified
to a limited extent at best, leaving the rest of the evaluation to subjective

judgment. That does not make such evaluations any less important to the T & E program and to the recommendations and conclusions that are to be drawn from the T & E program results. The TEMP and lower level T & E plans should describe such characteristics and how they will be evaluated.

Much has been learned regarding the more useful features of a TEMP. In Appendix C, we have expanded upon the guidelines described here, and have provided a notional TEMP for illustration purposes.

## 5.3 THE T & E EVENT PLAN

The test plan should be a stand-alone document that should be readily understood and interpreted both at the time the testing is conducted, and even years later when it might be used, for example, in the development of the next generation of the system or perhaps even in a court case. The following are the major items that should be addressed in a test plan:

(A) System description and configuration. Although the system description will appear elsewhere in requirements documents and specifications, it is important that it be repeated in each test plan, and if necessary expanded upon to make clear the reasons for this particular test event and the relative importance of the various test objectives. The system description describes how the system will be used – its operating profile. In addition to a system description, the specific configuration of the system (hardware, computer software, logistics support, interfaces, etc.) for this test event are described, and major differences between this configuration and that anticipated for the production baseline configuration are highlighted.

(B) Objectives of the test event. These should be clearly stated, and prioritized to the maximum extent possible as guidance to the test team to use if they need to quickly restructure the test procedures or consider suspending the tests because of emergent problems. Typical objectives might include:

- proving a concept
- ensuring safety
- ensuring adequate human factors engineering
- ensuring user requirements are met
- avoiding failures in service
- ensuring contract compliance
- verifying that fixes and enhancements work

- supporting investment decisions
- providing feedback to the designer
- confirming supportability
- validating models and simulations, and
- comparing systems.


(C)   Anticipated limitations to the scope of the event. In particular, limitations that might impact the ability to draw conclusions from the test results or to make recommendations on how to proceed need to be identified. While such limitations might be obvious to those close to the T & E planning, they may not be to others. An important part of the test report will be a description of which limitations actually did impact the T & E. Typical limitations are:

- cost
- safety
- capability of test instrumentation
- available time
- number and availability of test articles weather
- security
- anticipated usage profile and performance envelope.


(D)   Supporting organizations and their responsibilities. The roles, duties, and authority of the test directors, test conductors, system operators and maintainers, witnesses, technical support personnel, data takers, safety engineers, etc. need to be included to publicize them and ensure they are understood by all.

(E)   Conduct of the test. The specific step-by-step test procedures need to documented. If they are not included in the test plan itself, they need to be referenced in it. Test procedures show set-up conditions, station assignments, data sheets, and the acceptable range of results for passing each test. The objectives of the various parts of the test plan are identified. The test schedule, and the interdependencies among the test procedures (particularly which tests or parts of tests are prerequisites to others) are included. Data requirements and the need for special instrumentation are identified, or referenced if they are published elsewhere.

(F)   The analysis plan. This plan addresses the variables of the test, and the analysis to be conducted. Included in the analysis is the form of the initial data output, subsequent operations on the data to obtain inputs to a measure of effectiveness, and the definition of that measure.

(G)   Required reports. The various reports (partial, interim, quick-look, and final), their timing, and to whom each will be distributed are listed.

(H)   Security. Any particular procedures to protect test data from intentional or unintentional compromise are described.

(I)   Appendices. Operating profiles, test protocols, data sheets, instrumentation plans or safety procedures might be better added as appendices rather than in the main body of the plan.

## 5.4   CATEGORIZATION OF TEST PROBLEMS

For purposes of interim reporting as well as final analyses, it is useful to have an agreed-upon categorization for problems that may occur during the conduct of the T & E. The following is one possible categorization scheme:

(A)   computer software

   A1   computer software discrepancy

   A2   computer software reliability problem

(B)   hardware

   B1   item discrepancy

   B2   item missing

   B3   item physically damaged

   B4   item configuration not current or correct

(C)   other

   C1   out of tolerance

   C2   equipment interface problems

   C3   electromagnetic interference

   C4   out of alignment

   C5   miscellaneous

(D)   normal wear-and-tear and aging

(E)   reliability

    E1   design deficiency

    E2   manufacturing defect

    E3   installation problem

(F)   supportability

    F1   allocated spare not available

    F2   needed spare not allocated

    F3   spare is wrong part

(G)   maintainability

    G1   instructions inadequate

    G2   instructions missing

    G3   failure induced by maintenance procedures

    G4   periodicity of maintenance incorrect

    G5   access for maintenance inadequate

    G6   built-in-test capability inadequate

    G7   external test equipment not compatible

(H)   equipment not tested

    H1   not available

    H2   prerequisite conditions cannot be met

    H3   equipment in lay-up status

    H4   test personnel not available

    H5   documentation not available

(I)   support equipment

    I1   air-conditioning inadequate

    I2   cooling water inadequate

    I3   electrical power insufficient

    I4   test equipment design deficiency

        I5     test equipment out of calibration

        I6     test equipment not available

(J)   test services

        J1     test facility not available

        J2     other external resources not available

(K)  safety

        K1    system safety problem apparent

        K2    personnel safety problem apparent

(L)   personnel and training

        L1     training procedures inadequate

              L1a  training insufficient

              L1b  prescribed personnel qualifications inadequate

        L2     personnel not trained in prescribed procedures

(M)  unknown

## 5.5 THE T & E REPORT

The single most important document in the T & E program is the test report. It communicates the results of a T & E event – the very product of all of the planning, strategizing, coordinating, organizing, test engineering, modeling, simulating, measuring, data analysis, evaluating, and funding that have been described previously in this book. The results are the *raison d'etre* of the T & E program, but the report is the all-important medium through which these results will be understood.

    There are four important parts of a test report:

(A)   The Executive Summary. The Test Report will be read by a wide variety of people with varying degrees of understanding of and interest in the details of the T & E event. The large majority will be interested only in a summary description of what was tested and what the results suggest. Yet the report must capture in one place, for the record, much more information. The executive summary describes:

- the purpose of the test event
- how the test event fits into the overall development program
- a system description (including the configuration used for this test)
- a brief description of the tests performed
- a description of the limitations of the test – both those that were planned and those that had not been anticipated
- the significant test results
- the significant conclusions and recommendations

(B)  Description of the conduct of the tests. This includes:

- the objectives of the tests
- the test sequence and schedule
- test protocols, scenarios, usage profiles selected
- test environment
- personnel used and station assignments
- test facilities used
- test procedures followed (or reference to where they are documented)

(C)  Test results.

- summary of the data collected and the methods used for analysis.
- comparison of results with the objectives
- discussion and analysis of each significant failure
- special interest items and considerations

(D)  Conclusions and recommendations.

- Conclusions drawn (prioritized and related to test objectives and expected system performance)
- Recommendations for additional engineering and testing, and recommendations regarding that are to be based on the results of this test event.

How the test report will be received is so important that it is worth extra attention to ensure it is self-explanatory and is not likely to be misunderstood.

There are two techniques used in major programs that help achieve this. First, a report should be drafted even before the test event commences. While the actual data will not be available, using fictitious but realistic data will help achieve a smooth logic and traceability from the test purpose through test analysis to the actual conclusions. This effort usually helps identify potential deficiencies in the testing, data collection and analysis plans that could jeopardize the effectiveness of the test if they are not addressed. If the event could easily result in one of several different outcomes, a draft report for each outcome should be prepared. Secondly, various parts of the final report should be completed by the test team members while the test event is ongoing, obviating the need to try to recall or reconstruct facts after the entire event is completed.

This concurrent development of the test report helps achieve another very important objective: timeliness. For test reports that are to be the basis of programmatic decisions, the trade-off between report completeness and time-liness becomes one of the biggest problems for the test team. With a progressive report preparation approach as described above, timeliness can be achieved.

## 5.6   T & E REQUIREMENTS IN CONTRACTS

In general, the *test* requirements imposed on an organization which will be developing a subsystem or component should be at the same level as the *design* responsibilities imposed on that organization. Test requirements may be specified in several ways. For example, using some fairly simple products as examples, test requirements may be stated in terms of:

(A)   the required results; for example, 'the circuit breaker shall be tested to verify that it does not trip when subjected to shock factors identified in XXXX standard specification.'

(B)   operating environment; e.g., 'the binocular eyepiece shall be proven to operate at altitudes up to 10 000 feet above sea level'. Note that, as stated here, this does not necessarily require that the eyepiece be actually tested at 10 000 feet if it can be 'proven' to the satisfaction of the customer in other ways.

(C)   criteria for verifying compliance; e.g., 'the detector shall be inspected to ensure that it does not contain foreign matter, such as dust, dirt, fingerprints, or moisture that can be detected by visual examination.'

(D)   interface requirement; e.g., 'the shoes shall be of the following standard men's sizes: 9 through 13, in half-size increments.'

Such T & E requirements will cause the designer to demonstrate compliance with the design requirements for which he is accountable. However, in complex systems, additional T & E will be usually necessary to help ensure that overall system effectiveness in the planned environment is achieved. While this T & E may be the responsibility of the next higher level of the program organization, the developers and testers of the subsystems or components should be participants in those tests to ensure that questions about interface designs or the earlier lower-level T & E can be correctly and expeditiously answered when problems arise.

## 5.7 SUMMARY

Written documentation of the planning and of the results of the T & E program is but one way in which understandings are communicated and misunderstandings are identified early among the many parties involved in the T & E program. In addition, documentation is a very necessary activity of which the timeliness and accuracy are integral to the adequate timeliness and accuracy of the entire development program.

## REFERENCE

USDOD (1991) *Defense Acquisition Management Documentation and Reports*. US Department of Defense Manual DoD 5000.2-M, February.

# 6

# T & E Techniques

*There are no shortcuts to any place worth going*

Beverly Sills

Test and evaluation has come to encompass many different types of systems, technologies, and disciplines. This chapter describes the methods used for specialized categories of T & E. It also describes the role of modeling and simulation in T & E programs for complex system developments.

## 6.1  T & E OF COMPUTER SOFTWARE

Today, computers control electrical power plants, they track our money, they control the timing and fuel injection in our automobile engines, and they navigate most of our commercial aircraft. Nearly all major operator-controlled systems are computer software-intensive. In addition, the complexity of computer software is dramatically increasing. It has been estimated that between now and the year 2010, there will be a 10% increase each year in the number of functions previously handled by hardware being transferred to software. Not surprisingly, a significant part of the cost of the development of the computer software is for testing: testing throughout the development in order to identify errors as early as possible, as well as for acceptance testing. Historically, a guideline for the percentage of program cost was 40%, as in the '40–20–40 rule' depicted in Figure 6.1 (Pressman 1987). However, many companies today estimate that the typical percentage for testing is now 50%. The cost of fixing errors – particularly software errors – compels the finding and fixing of such errors as early as possible in the development process. Figure 6.2 shows how such costs typically increase the further into a program they are incurred. In the past few decades, the technical complexity of the designs of new systems has generally outpaced the capabilities available to test them, creating a major challenge for the T & E program.

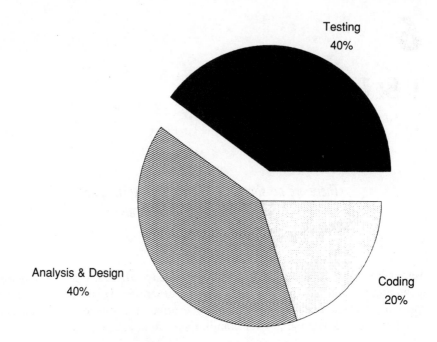

**Figure 6.1**
Cost distribution for computer software

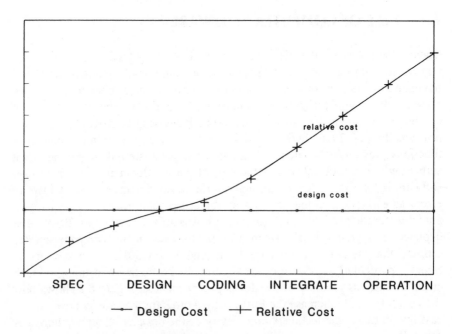

**Figure 6.2**
Relative cost of error correction

## 6.1.1   Testing Techniques

Software has 'bugs'. That is a fact of life. Today's large sophisticated computer programs perform so many functions and have so many logic paths that the bugs can go undetected through the *entire life* of the software. The challenge for the developer is to adopt a disciplined, well structured development and T & E program that will provide reasonable assurance that the potentially critical bugs – those that would jeopardize system operation or the safety of personnel – have been identified and corrected before the software is put into service. Adopting such an approach requires not just an emphasis on software quality standards and techniques, but a truly pervasive commitment to finding and fixing problems as early as possible.

There are several commonly used classical testing techniques for software, listed in Figure 6.3, each of which has its strengths and weaknesses. White box testing, useful on older, smaller programs, and perhaps in some small modules of today's larger programs, consists of thorough ('100%') testing of every function and every path. It is based on good access to the program logic and input/output domain information. It attempts to execute most execution paths with representative data. Black box testing, on the other hand, assumes very limited access to the program. It involves testing the performance (output) of the total system ('black box') based on a selective set of stimuli (inputs). A major weakness in black box testing is that it assumes that the specification is correct. Black box testing became particularly popular in the 1980s as a way to save test time during the development of ever-increasingly larger programs, even when the functions and logic were well known. However, it was frequently used to the exclusion of white box testing, allowing some design problems to go undetected until late in development. Today, layered testing of computer software, with a balanced mix of black and white box testing, is seen as the best approach.

- WHITE BOX TESTING

- BLACK BOX TESTING

- STATIC TESTING

- DYNAMIC TESTING
  - mutation testing
  - domain testing
  - partition analysis
  - functional testing

- BETA SITE TESTING

**Figure 6.3**
Software testing techniques

A frequent mistake in scoping software T & E is to assume that exhaustive testing can be done. Usually it cannot be. For even small software programs, the number of logical paths can be very large. For example, consider the flow chart in Figure 6.4 (Pressman 1987). The design illustrated might correspond to a 100-line Pascal program with a single loop that may be executed no more than 20 times. There are approximately 100 million possible paths that may be executed! If there were a test processor (there isn't) that could develop a test case, execute it, and evaluate the results in one millisecond, and it worked 24 hours a day and 365 days a year, the processor would need to work for 3170 years to completely test the program represented in the flow chart. Another dimension of software test strategies is dynamic analyses. This is a family of techniques such as mutation testing, path and domain analysis, path analysis, and partition analysis that can be drawn upon to help test today's software development programs.

Because so much system functionality of today's systems resides in the computer software and because software development has become so competitive, most software developments are user-focused from the beginning. Software developers frequently establish useability laboratories to help evaluate early the tasks the users perform with the software and to take empirical measurements, feeding the results back into changes to the design. In International Business Machine's (IBM's) Atlanta Usability Laboratory, each

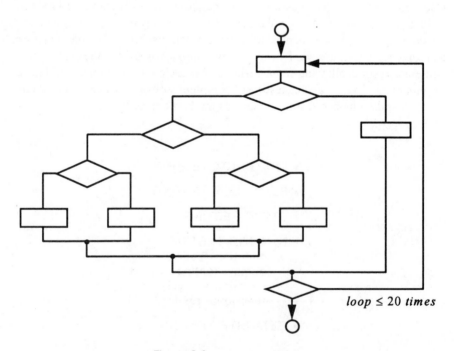

*loop ≤ 20 times*

**Figure 6.4**
Problems with exhaustive testing

of the lab's modules has three rooms: a product evaluator's area (typical users), a control room for the test directors, and an observation room for other interested parties. Cameras can record the evaluator's keyboard, display screen, or even facial expressions. The module can be set up to simulate a number of typical user work environments such as computer integrated manufacturing, desktop publishing, or medical and personnel management products (Fath, Mann and Holzman 1994).

Another approach that has become common in large computer program developments is 'beta site' testing. During this phase of development, versions of the software are installed at typical user's sites for operational T & E. Before Microsoft Corporation introduced its 'Windows 95' program, it used an army of over 400,000 beta testers to search out the bugs. Software beta testing for smaller programs is spawning businesses such as the 'Bug Police', a company started in San Francisco in the mid-1990s to provide such testers (USA Today 1995). It is not just the program developers that are subjecting their products to individual consumers before release. After serious problems were found with the computer processing chip in IBM's Pentium computers when they were introduced in 1995, the chip maker, Intel, decided to send its next generation arithmetic processing chip not just to its immediate customers, the computer manufacturers, for testing prior to release, but also to individual beta testers.

## 6.1.2   Metrics

In the 1960s and 1970s computer software was thought of as a product in itself, developed and tested by specialists and installed in the system after the hardware and software had completed development. In the 1980s, as more and more system functions gravitated from residing in the hardware to residing in the software, the hardware and software were co-developed; testing of both together was relied upon to uncover deficiencies and to verify proper system performance. In the 1990s, it was recognized that today's systems are too large and complex for T & E to demonstrate their full ranges of capabilities. Besides, the full system T & E that is conducted occurs too late in the program to adequately reduces risks. So during development, T & E organizations now look to various metrics to provide assessments of how likely the particular development *process* is to produce quality software. Among the more common and popular metrics are: status of funding, size of the computer program (including the ratio of new code versus reapplied code), volatility of the requirements, use of computer resources such as processor memory, processor throughput (processing time) and number of input/output channels), and design complexity. The more common metrics for assessing technical progress are: volatility of requirements, adherence to schedule, status of testing, and the status of Problem Reports.

(A)  Volatility of requirements. Changes in the performance requirements, or changes in the developer's interpretation of those requirements, are probably the single largest factor causing software development cost and schedule overruns. Two metrics are useful: the number of approved changes to requirements, and the number of design issues that cause them. Regarding the first, the plot of cumulative changes should rise more steeply prior to the first major design review and level off after the final design review. Figure 6.5 is an example showing a track of a program approaching its final design review. Note that the changes after the initial design review may result in a schedule slip for final design review because some software designs may have to be changed. Schedule slips are a risk to T & E because managers frequently compress T & E and therefore increase risk in order to accommodate the slippage. Regarding the second metric, a plot of time versus the number of design issues for a healthy program would be expected to show a rise at each major design review and a high

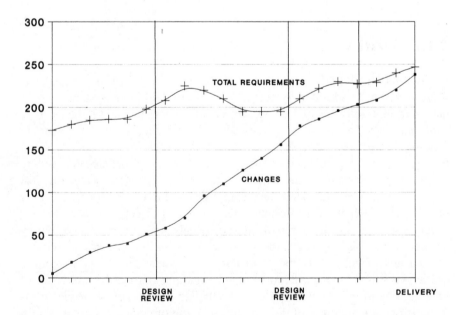

**Figure 6.5**
Number of requirements

**OPEN DESIGN ISSUES**

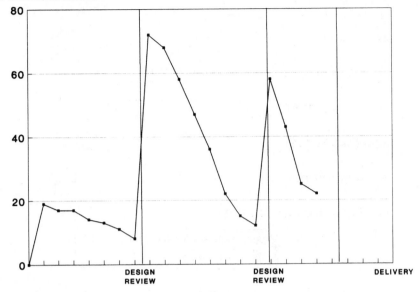

**Figure 6.6**
Counting design issues

rate of tapering off as those issues are quickly closed out. The number of open design issues would be expected to decrease as the program proceeds. Figure 6.6 is an example.

(B) Adherence to schedule. As discussed before, schedule slippage frequently causes compression of available test time, in addition to its other undesirable effects. In the example in Figure 6.7, the development of the Program Performance Specification took longer than expected. This forced a delay in the initial design review. The chart shows the anticipated effect on the remaining major activities in the program. In this revised schedule, program acceptance testing is anticipated to require more time, but the subsequent integration testing is holding to its original schedule. This could be an indicator of overly optimistic scheduling. Management needs to concentrate substantial effort to ensure that realistic schedules are maintained.

(C) Status of testing. Two closely related metrics are the status of testing and the status of Problem Reports. Testing can be tracked by plotting the number of planned test completions versus the number of actual test completions. In Figure 6.8, some units that have started formal

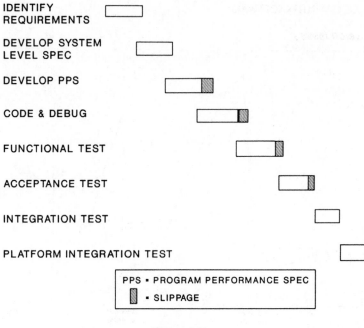

**Figure 6.7**
Test schedule

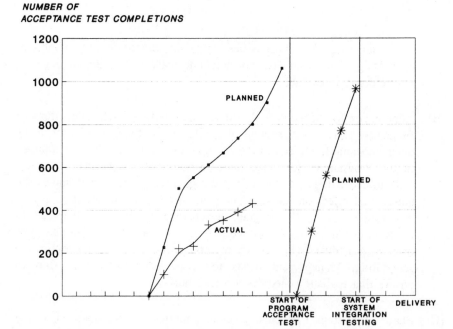

**Figure 6.8**
Number of test completions

acceptance testing (about the time of the Test Readiness Review) have not completed, i.e., have failed. It would appear that the start of program acceptance testing will be delayed. The reasons for the failed tests could be late changes in requirements or design deficiencies. If the software modifications made as a result of these failures are significant, they could invalidate the results of some previously successful tests, causing further test slippage to allow for those tests to be repeated.

(D)   Status of Problem Reports. The number of Problem Reports per 1000 Source Lines of Code is an indication of both testing adequacy and of code quality. Typically, the range will be between 5 and 30 Problem Reports per 1000 Source Lines of Code, with 10 to 20 a safer range. The norm from several sources averages about 16 (Beizer 1983). Too few Problem Reports may indicate poor testing and too many may indicate poor code quality. In the example in Figure 6.9, note that the Problem Report density (total number of open Problem Reports) ranges between 8 and 20, which is within the normal range. A better indicator of code quality is shown by the chart in Figure 6.10, an ideal case. Each Problem Report is categorized into one of five priorities according to its impact on system operations: (1) prevents accomplishment of a critical operation; (2) adversely impacts a critical operation and no work-around solution is known; (3) adversely impacts a critical operation, but a work-around is available; (4) results in user or operator inconvenience but does not affect a critical system capability; (5) all others (USDoD 1995). Each priority is assigned a weight, and the number of open Problem Reports multiplied by their respective weights is plotted over time.

The metrics described above are interrelated, and need to be interpreted as such. No single metric is meaningful in determining whether the program is on track. As reflected in Figure 6.11, all are needed to get a good indication of progress towards achieving a quality software delivery.

## 6.1.3   Software Test Tools

The software developers now believe that even extensive beta testing cannot be counted on to uncover sufficient errors. The opening of the new Denver International Airport was delayed four times, from October 1993 to February 1995, due in large part to software and mechanical problems in its computerized baggage handling system. This system is the most sophisticated of any airport in the world, involving 22 miles of track and 6 miles of conveyor belts. A new generation of automated software tools has been fielded to test large software-intensive systems and networks. These tools aim to shorten the time to test (and the associated costs) or to allow more extensive testing, or both.

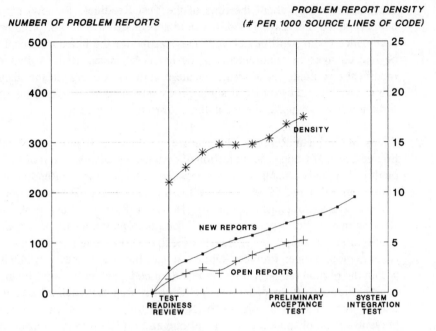

**Figure 6.9**
Number of problem reports

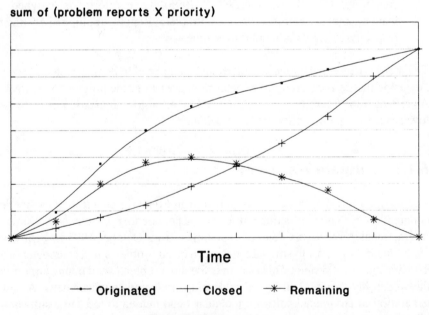

**Figure 6.10**
Priority of problem reports

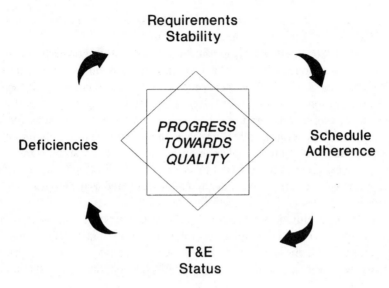

**Figure 6.11**
Interrelationship of the software metrics

## 6.1.4   Management Techniques

The well established desktop reviews of coding by the software developer and peer reviews by other engineers on the team remain effective ways of uncovering software problems early. At the other end of the spectrum, Independent Verification and Validation (IV & V) is a more formal way of selectively obtaining objective scrutiny of the software. IV & V was born during the early days of the space and missile programs, because of the need for the software to work properly the first time it was used operationally. An IV & V team, organizationally separate from the developers on the program team, independently tests the software, assesses whether it satisfies all system-level requirements, and perhaps even develops a separate software package for a critical item.

## 6.2   T & E OF RELIABILITY

The extensive work in reliability engineering in the last several decades includes mathematical modeling, and advancements in development of design techniques for achieving the required levels of reliability, maintainability, and availability (RM & A) in complex systems. The modeling allows planners to accurately predict not only the final reliability, maintainability and availability of the end product, but also at what rate the values of those parameters may well grow during the course of the development program.

In the late 1980s, there was a significant paradigm shift. Formerly, full-scale system demonstrations at the end of a development program were relied upon as the primary means by which RM & A requirements were enforced. Instead of such demonstrations, system RM & A is now tested, using operational profiles of the system, *throughout* the program. The well accepted design practices for high reliability – such as derating, stress analysis, thermal analysis, and failure mode and effects analysis – are complemented by an iterative test-analyze-and-fix (TAAF) process. Sometimes referred to as a reliability development test (RDT), development systems are tested under actual or simulated usage environments to uncover deficiencies and to provide information on failure modes. The objective of the RDT is to provide a basis for developing, verifying, and incorporating corrections early. The RDT emphasizes reliability growth rather than numerical measurements, which were predominant a decade ago when the primary reliability test was a qualification (pass/fail) test conducted at the end of development.

Martin Meth, from the staff of the US Office of Secretary of Defense, cites several 'rules of thumb' in reliability T & E that are based on empirical observation, or can be first-order approximations of complex mathematical models, but do not have a statistical basis (Meth 1994):

*The 'Rule of Ten': there will be a high confidence with the observed mean time between failure (MTBF) if the test time is ten times the observed value.*

The rule appears to work because the test time is relatively large compared with the observed reliability value. High confidence is considered about 90%. Using this rule, one can estimate the test time needed to prove out a predicted reliability. Interestingly, the rule will reduce the producer and consumer risks to reasonable levels.

*In planning reliability growth tests, allocate from 1000 to 2000 hours of testing with at least two units for a period from one to two years.*

In part, this rule may be considered an offshoot of the Rule of Ten. Enough test time needs to be scheduled to find failure modes that should not be occurring within a reasonable operating time. Most complex subsystems using reliability growth testing have mean time between failure values of between 5 and 30 hours. Hence, the Rule of Ten would suggest 500 to 3000 hours. However, resource constraints usually limit the investment in reliability growth. Experience with the numerous reliability growth tests shows that, if they are properly constructed, unexpected failure modes will show up early in the first several hundred hours of testing. The remainder of the test time is used to verify the corrections. The number of components and calendar estimate are also based on experience. Generally it takes from 1 to 5 months to determine failure cause and develop a correction.

*To verify correction of a failure mode, the test time should be at least three times the experienced failure frequency of the failure mode being corrected.*

The idea behind this rule is that the failure mode has likely been corrected because the test time is large enough. The key here is *likely* as opposed to '*highly likely*'. Usually, testing three times the expected reliability value will result in a confidence estimate of better than 50%, but not nearly as good as a 90% confidence estimate. The rule provides a necessary but not sufficient condition. But the rule reflects the practical constraints for economically proving out corrections. In fact, one of the first rules that every reliability person learns is that there is never enough money for a reliability program.

*On the average, the effectiveness of design correction to improve reliability will be 70%.*

Here is another rule that can save money. This rule is based on experience. No matter how hard one tries, there always seems to be some aspect of design that is overlooked.

*The predicted reliability should exceed the contractual reliability by at least 25%.*

When making predictions, people tend to be optimistic. A prediction that is close to the desired value is a warning sign that there is little margin for error. The real lesson is to pay attention to design margins. In a particular program, this rule of thumb should not be visible, lest it become a self-fulfilling prophecy, with predictions becoming automatically 25% greater than the contractual reliability.

*In reliability growth testing of new equipment, the ratio of design pattern failures to workmanship failures should change from about a 60:40 ratio to 20:80 over the test period.*

Reliability growth testing has as a principle objective the correction of design failures through a test-and-fix process. If the test activity is working as anticipated, then there should be fewer design pattern failures as the testing progresses. If a ratio change of this magnitude is not observed, a lengthened reliability test program should be planned.

*If a reliability growth curve for a new equipment follows a straight line (a log-log plot), then the test environment is not adequate.*

A new equipment design generally has thousands of failure modes. It is unlikely that the design, even using the most up-to-date design rules, will be able to account for all environments. Unanticipated failure modes are likely and usually occur in 'bunches'. The usual shape of a reliability growth curve includes a steep downward (negative) slope followed by a steep upward (positive) slope. Not seeing such a pattern is usually an indication that the growth process is not taking place.

*Thermal soaking or dwell is less effective than thermal cycling in uncovering unanticipated failure modes.*

Thermal cycling is a form of accelerating equipment life. Usually 15 to 20 thermal cycles are adequate to work the hardware to a point where mechanical stress will cause failures. Similar experience with a broad vibration spectrum has not highlighted any repeatable pattern, although many believe that low-frequency vibration is generally more damaging. The bottom line is that being passive is not the role of the reliability engineer.

## 6.3 ENVIRONMENTAL T & E

Systems and products must be tested to ensure that they will perform adequately and not become unsafe in the environmental conditions they will be subjected to, and not contribute to unsafe or bad environmental conditions in other systems. Figure 6.12 shows some of the more common environmental areas that need to be considered in planning a system T & E program. There are many test standards – international, national, government and industry – that are of use in planning environmental T & E for complex systems and subsystems. See section 4.3.1 for a discussion of T & E standards in general. There are also many test stands and chambers available, in both independent test laboratories and in company laboratories, for inducing the large variations of parameters needed for system and component testing in development programs. Information on such facilities is readily available in catalogs, both written and in on-line computer databases, as well as through individual Home Pages on the World Wide Web.

| NATURAL | INDUCED |
|---|---|
| temperature | thermal shock |
| humidity | temperature |
| rain, snow, sleet, hail, ice | pressure |
| dust, sand, gravel | acceleration |
| salt sea spray | vibration |
| solar radiation and darkness | shock |
| lighting | acoustics |
| insects, birds, small animals | corrosive elements |
| biological growth | electromagnetic interference |
|  | radiation |

**Figure 6.12**
Environmental T & E

# 6.4   T & E OF HUMAN ENGINEERING

The systematic design and development of the human components of systems is something of relatively recent vintage. Work in the disciplines of industrial engineering and psychology since World War II has caused a recognition that systems engineers can and should design to optimize the man–machine interface in today's complex systems. Through advances in what is now called artificial intelligence, system developers have tried to limit the variability of overall system performance by limiting the variability of the human performance, through automation of as many functions as possible.

A good developmental T & E program will have a series of tests prepared specifically to evaluate human performance, rather than counting on human factors problems to be identified only during the hardware/software T & E. Historically, the primary method of conducting an evaluation of human factors was through interviews of people who had actually used the system. This was valuable in determining what was wrong with the system design, but not necessarily what was *right*, or what was satisfactory but should have been improved. Now that there are good guidelines on how to design-in human factors as an integral part of the systems engineering effort, desktop reviews, checklists, and activity analyses by human factors experts provide many more opportunities to improve the human part of a system design earlier, and to have a greater impact on it. Also, the increased use of modeling and simulation in designs allows an early evaluation of human factors, well before any hardware is actually built. System integration test facilities (Section 4.3) also provide an opportunity for T & E of human factors before final assembly and mass production begin. Once a full system is available (e.g. preproduction prototype or early production unit), human factors T & E can focus on the integration of human, hardware, and computer software objectives:

- to confirm that the capabilities of the total system are not unnecessarily limited by the operators' capabilities;

- to verify that the system can meet its human performance requirements consistently and reliably. Human performance requirements could include items such as operability, habitability, maintainability, transportability, and portability by the person who will use it; and

- to determine the adequacy of operational and maintenance procedures (including diagnostic tools), technical manuals for the users, and training for the users.

T & E procedures to evaluate human factors engineering covers topic such as: the adequacy of lighting, noise/hearing protection, temperature/humidity/ventilation, visibility, speech intelligibility, work space and anthropometries, force/torque demands, demands on dexterity, panel commonality, maintainability, individual performance, team performance, error likelihood, training, task complexity, and workload.

## 6.5  TESTING FOR SAFETY

In *Escola* vs. *Coca Cola Bottling Company of Fresno*, Justice Traynor pointed out in 1944 that 'the manufacturer must know the product is fit, or take the consequences, if it proves destructive.' He also said, 'The only way a manufacturer can know his product is fit is by analysis and test.' (Hammer 1980). The well known safety engineering techniques include fail-safe designs, interlocks, damage minimization and containment, and isolation. Before a product is released, analyses (in particular, failure modes and effects analyses and fault tree analyses), demonstrations and even simulations can be valuable methods of assessing safety and uncovering potential hazards. When full system prototypes and early production units are available, specific safety tests should be conducted. The tests should be well planned to stress safety areas that might not have been stressed in earlier testing, and also to validate the analyses that have previously been conducted. These tests should be done with the full participation of a qualified safety engineer.

## 6.6  DESIGN-LIMIT TESTING

Design-limit tests are those tests used to gain an engineering understanding of the design and to suggest aspects of the design that need improvement. The end objective is to ensure that the system will provide adequate performance characteristics when exposed to environmental conditions expected at the extremes of the operating envelope. Usage profile testing is essential to sound design limit testing. Studies should be conducted early to accurately characterize the usage profile, the most stressful system modes and the 'worst case' environments. This also includes subsystems which may have increased stress due to amplification factors and heating problems or decreased stress due to shock and vibration isolation or efficient cooling mechanisms.

## 6.7 LIFE TESTING

The ability of the system to withstand long-term exposure to the operating environment is essential. Life testing, conducted to demonstrate this ability, is often run too late to impact the design. Instead of being conducted at the end of development, it should be conducted throughout, integrated with other parts of the subsystem and system T & E programs. Detailed analyses of the life characteristics should be made concurrent with the initial design. Aging failure data should then be collected from existing equipment, or equipment similar to that being designed. For these analyses, a judicious life test program, integrated with other T & E events, can meaningfully be planned.

## 6.8 INTEROPERABILITY TESTING

For many types of complex system, interoperability is a major design objective, and therefore a significant T & E objective. Simply defined, interoperability is the ability of systems to provide services to or accept services from other systems and to use the services so exchanged to operate effectively together. Electronics-hardware miniaturization and high-speed computing have made dramatic increases in interoperability possible. In this Information Age in which we live, our access to and use of communications and database suprasystems are commonplace. The military programs have paved the way in interoperability design and testing, where for decades the reaction times necessary to counter the threat systems have made maximum interoperability mandatory. The primary driver in interoperability in military systems design has been between the command and control functions and the various sensors and weapons in a given system. However, as shown in Figure 6.13, interoperability now involves the automated integration and correlation of inputs from a variety of platforms (space-based, aircraft, surface and subsurface) and multiple systems (sensors, command and control, communications, weapons control, launchers, weapons, and support systems). Interoperability relates not only to inter-Service and joint Service compatibility, but also compatibility with other nations' military systems.

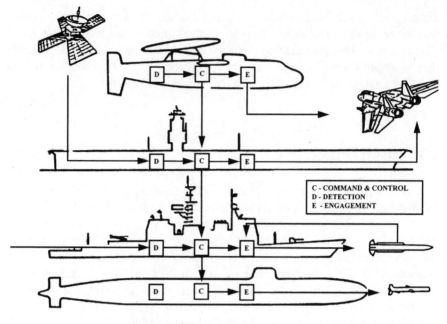

**Figure 6.13**
Interoperability among military systems

Some major advancements have been made in interoperability T & E capabilities. Communications and computer technology, including satellite linkages, allows real-time electronic connectivity between facilities such that the actual geographic location of each of the participants can be irrelevant. Much of the initial work has involved federating and integrating systems that are already in service. Interfacing systems have evolved that provide quick and accurate translation of information between such systems. In addition, new standards have been developed for both interoperability design and testing.

## 6.9 'LIVE-FIRE' TESTING

In 1986, the US Congress amended Public Law by adding requirements for 'live-fire' testing in the assessment of the vulnerability of manned military platforms and the lethality of weapons (US Code 1986). In a reaction to what Congress felt were shortcomings in the scope of such testing in the past, the law now stipulates that covered systems (such as tanks, aircraft, and ships) are to have survivability testing completed prior to major programs proceeding beyond low-rate initial production. Further, at the completion of such testing on each program, the

Secretary of Defense is to submit a report to Congress. When feasible, live-fire testing includes firing of live ordnance against combat-loaded US systems to test their vulnerability, as well as the firing of US munitions at operational, combat-loaded threat targets (or surrogates) to test their lethality.

Modeling and simulation efforts are essential elements of 'live-fire'. They have historically been major tools used in the assessments of survivability and lethality, and remain so even though additional full-scale testing is now being conducted. Modeling and simulation are not just critical to extrapolating test results from a necessarily limited number of shots. They are also used during test planning to identify the scenarios and shot lines selected for live firings.

## 6.10  SYSTEM TESTABILITY

The ability of a system to be tested not only during development but also during production and while it is in-service – both effectively and efficiently – must be *designed into* a system, early. The common design techniques shown in Figure 6.14 are proven and well known (USDoD 1985), but incorporating them into a development program requires a strong commitment among the entire program team. Only when the importance of testability is fully recognized from the beginning of development are the design techniques adequately accommodated. During early design, the performance characteristics to be measured during production and in-service testing must be identified, data on them must be collected during development, ready access to those parameters must be designed, and the design must enable rapid and accurate assessment of the status to the lowest repairable element

> specific performance characteristics must be identified

> ready access designed in

> status assessment to lowest repairable level when
    deployed

> physical & electrical partitioning

> system level & item level built-in test (BIT) throughout

> balance between measurement manually, by BIT, and by
    external Automatic Test Equipment (ATE)

> ATE designed concurrently with the system

**Figure 6.14**
Techniques for designing for testability

when the system is in operation. A design that is easily and completely testable, without disassembly, adjustments, special environmental conditioning, or external equipment or stimuli for monitoring of responses, is amenable to economic production. Designing for testability should include physical and electrical partitioning; system-level and item-level built-in-test (BIT) capability distributed throughout the system; a well engineered balance between characteristics to be measured manually, those by BIT, and those requiring external Automatic Test Equipment (ATE); maximum compatibility between the system and the ATE; and good test control and access. The design trade-offs between BIT, external test equipment, and manual testing should be done early so that the ATE is designed concurrently with the prime equipment, and is available for testing with the system during the later stages of development. A frequent misconception is that ingenuity in the design of test equipment can compensate for deficiencies in the testability of prime hardware. In reality, not much can be done to 'add on' testability if provisions for it were not made in the original design. No amount of breakout boxes and extender cards can compensate for poor testability design. A design that is fully maintainable is quite likely highly producible. If fault isolation needs are met, if access to signal flow is provided, if modularity of function is provided, if knowledge of proper performance can be determined without introducing external stimuli or monitoring requirements, and if alignment and adjustments are minimized, the production and in-service maintainability needs regarding testing will be met.

Testability itself can be evaluated during system design. The coverage of fault detection across components and functions can be measured. The accuracy of fault detection and fault isolation as well as the number and frequency of false alarms can be estimated, set as an objective, and then tested. There are also automated testability analysis tools available which provide testability figures of merit for the various design configurations during the development program.

## 6.11   MODELING AND SIMULATION (M & S)

A model is a representation of a system that replicates part of its form, fit, function, or a mix of the three, in order to predict how the system might perform or survive under various conditions or in a range of environments. A simulation is a method for exercising the model. Simulation is closely akin to T & E in that both measure system performance. In T & E, performance is measured on an actual system, whereas in simulations, performance is measured, at least partially, in models of the system. Simulation of a system can be physical, can be solely computer based, or can involve actual components of the system ('hardware-in-the-loop') and/or user personnel ('man-in-the-loop') mixed with models of other components. Typically, simulations early in a development program are primarily computer based, and later they begin to include actual system components and users.

## 6.11.1   Physical Models

Physical models have always been used to support T & E programs. For example, plywood mock-ups are used to verify that planned hardware placements in tight spaces such as ships, aircraft, and automobiles optimize the ability of the operators to perform their tasks, and allow for ease of maintenance and parts replacements. Another example is the use of Calspan Corporation's Atmospheric Simulation Facility in Buffalo, New York, pictured in Figure 6.15. This facility is used to optimize the design of snow skis, particularly for ski-jumping. During the tests, a ski-jumper is suspended in chest and thigh harnesses like a marionette, while his skis are attached to struts in the floor of the tunnel. The result is a state of weightlessness. This allows the test conductors to measure the jumpers's lift and drag forces by monitoring the strain on torsion bars attached to the

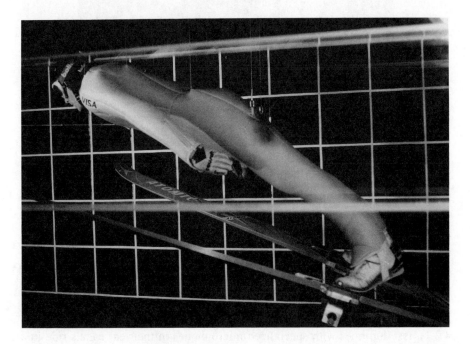

**Figure 6.15**
Calspan's atmospheric simulation facility

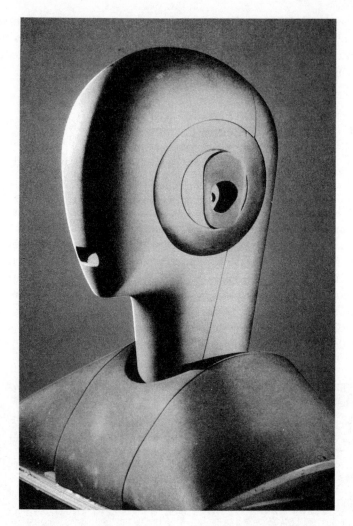

**Figure 6.16**
Chrysler's NVH testing dummy

harness. During the half-hour sessions in the tunnel, the jumpers can make countless balance adjustments, in contrast to actual conditions on the ski slopes where the take-off and flight positions of a jumper are of very short duration, allowing almost no adjustments (Holden, Girardi and Florio 1991). Automobile manufacturers use articulated dummies for a variety of purposes. At Chrysler's Noise, Vibration and Harshness (NVH) Laboratory (see Section 4.2.1), test dummies with specialized microphones in their ear canals ride in a vehicle as it travels on the special chassis rolls, monitoring sounds, as shown in Figure 6.16.

## Case example: crash dummies

With the enactment in the US of the National Highway Traffic Safety Act of 1966, automobile crash testing and the evaluation of crash dummies took off. Each of the major automakers as well as organizations such as the Insurance Institute for Highway Safety (see Section 1.3) make substantial investments in developing and improving crash testing and crash dummies. Representing a range of human sizes and conditions – from infant to adult, from small female adult to large male, and even a pregnant female – the dummies are complex instruments that have to be calibrated in a laboratory to ensure that they respond the way a human body would in a real crash. A single instrumented dummy can cost $100,000. They can be instrumented to measure a variety of forces, such as frontal impact, acceleration, side impact, velocity, damage assessment, and projectile penetration.

Every year, General Motors alone conducts about 500 car collision tests with such dummies (Wolkomir 1995). One of them, the Side Impact Dummy (SID) developed by the University of Michigan, provides 31 continuous measurements during a test. SID was adapted from the Hybrid II 50th percentile Male Test Dummy to provide human-like acceleration responses in the lateral direction when testing automotive side impact crashworthiness. In the latest generation dummy, Hybrid III, which is still evolving, devices measure head impact, neck loading, chest acceleration, chest deflection, femur loading, tibia and fibula loading and pelvic acceleration.

Calibration is conducted in a climate-controlled lab because temperature and humidity can affect the results. Dummies are stored in a climate-controlled room to maintain adjustments until vehicle preparation is complete.

First Technology Safety Systems in Plymouth, Michigan has 90 workers who manufacture most of the world's crash-test dummies, about 300 per year. They have customers in virtually every country in the world. They also provide for their customers their services for rebuilding, spare parts, and recalibration for dummies that are damaged. One of the company's slogans is, 'People can't be replaced. So we make replacements for people.' The company goes to great lengths to ensure the biofidelity of the dummies. The porosity of the aluminum bones is checked by X-ray to make sure they do not break prematurely. First Technology is now developing a facial skin that will show lacerations and abrasions. The company manufactures six sizes of the Hybrid III as well as 6, 12, and 18-month old infant dummies called 'CRABI' (for Child Restraint Air Bag Interaction). Figure 6.17 shows First Technology's complete 'dummy family'.

Although auto crash tests provide the primary market for such highly instrumented dummies, they also have other uses. The US Defense Department uses dummies to test, for example, the impact of a wound and the ability of the armor on an Army truck to protect the occupants from the detonation of a land mine. The Federal Aviation Administration uses dummies to evaluate design and installation of seats in aircraft.

Engineers are now working on electronic models of dummies for use in computer-simulated crashes that may eventually remove the need for many of these dummies. First Technology is now working with Ove Arup & Partners of England on finite-element models of the complete line of crash dummies they manufacture.

**Figure 6.17**
First Technology Safety System's 'dummy family'

## 6.11.2   Computer Based Models

Computer based models and simulations are becoming much more sophisticated and commonplace. More and more, computer technology is providing insights into how our world works, from glimpses of the body's molecular structure to understanding weather patterns. The ability to recreate the world around us, particularly via computer M & S, is providing more control over both natural and man-made forces, and offering an ever-increasing number of potentially valuable applications, from training to developing drugs that cure life-threatening illnesses. Thanks to the availability of relatively inexpensive computers that store and process information at very high levels, complex M & S, once only the domain of governments and academia, are becoming increasingly available to a host of industries. Improvements in simulation software have dramatically reduced the costs and time to develop models. In addition, the availability of graphic animation has resulted in a greater understanding and use of M & S by managers, engineers, designers and testers.

## 6.11.3   M & S in Support of T & E

The most common uses of M & S in support of T & E are:

- to assist in identifying critical test objectives

- to identify important parameters for T & E and the interrelationships among them early

- to augment, extend and enhance system T & E results where it is impossible or impractical to physically test

- to represent the whole system when components are not available

- to represent the environment for the system for evaluating compatibility and interoperability where actual T & E is impossible or impractical.

For example, biochemists at the University of North Carolina at Chapel Hill are using M & S to manipulate the molecule of a potential anti-cancer drug to a so-called receptor site on a protein to test the drug's effectiveness. Companies like DuPont Co. simulate disposable diapers in order to test ways to make them more absorbent. Caterpillar Inc. uses digital prototypes to test designs for earth-moving equipment. M & S has become not just a practicality, but a necessity. To show the importance M & S plays in today's complex T & E programs, we add a loop to the T & E engineering process described in Chapter 4 and depicted in Figure 4.1. The resulting process chart is Figure 6.18. The M & S loop depicts in particular the roles of M & S in supplementing the T & E to be conducted and in supporting the analyses of the results of T & E.

## 6.11.4   Verification, Validation and Accreditation

These three terms are coming into common usage to reflect the different levels of formality in determining the credibility of a model. A model is said to be 'verified' if it has been determined that it satisfies the system developer's conceptual description of the system. It is 'validated' if there is generally acceptable evidence that it is a reasonable abstraction of the 'real world'. It is 'accredited' if it has been certified, usually by an independent expert, that it is an

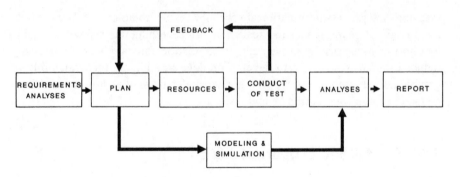

**Figure 6.18**
T & E engineering process with M & S

acceptable replication of the system for a specific application, e.g. for use in achieving the objectives of a particular T & E event.

## 6.11.5   Key Steps in Developing a Computer Based Model

Figure 6.19 shows the major steps that need to be taken for a successful M & S effort (Law and McComas 1990). First, the objectives of the M & S effort must be identified in order to help focus everyone's efforts and to help prioritize the work. They should be as specific as possible. An M & S effort to feed into a one time investment decision for a new system would require a markedly different approach from one that would be used for weekly production scheduling. Who will actually operate the M & S and who will determine its outputs should be part of this objective setting. The objectives should also include identifying what parameters are to be measured through the simulation.

Then, the M & S developer collects information on the system operation and its control logic to define the model. This should be collected not only from system publications but through contact with actual and prospective system users. If possible, data should be collected to identify model parameters and input probability distributions (e.g. for machine operators and repair times). In general, each source of system randomness should be represented by an appropriate probability distribution (not just the mean) in the model. Furthermore, the validity of each distribution should be evaluated by using statistical analyses. This information should be captured in an 'assumptions document.' The objectives, the availability of 'real world' data, credibility concerns, and resource constraints will bound and help define the level of detail of the M & S effort.

Before coding begins, the M & S developer should do a deliberate review with

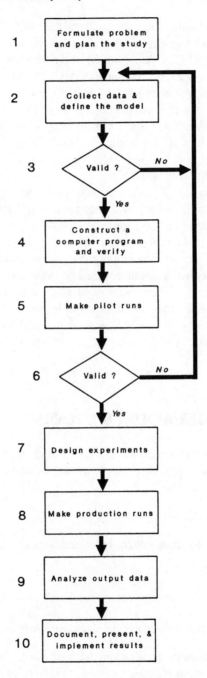

**Figure 6.19**
Steps in a modeling and simulation effort. Reprinted with the permission of the Institute of Industrial Engineers, 25 Technology Park/Atlanta, Norcross, GA 30092, (770) 449-0461. Copyright © 1990.

the program manager and others who will use the results of the M & S in order to verify that the model's assumptions are correct, complete, and consistent.

For actually developing the model, there are many different types of simulation program already available today. Usually, these can provide a good foundation for a system simulation effort, but some additional programming will be necessary to meet the objectives of the effort. Another capability more readily available today that can help with selling the value of the M & S effort as well as help with interpreting its results is animation.

Once the program is developed, 'pilot' runs are made for validation purposes. Output data can be compared against what was expected, and sensitivity analyses can be conducted to determine what model aspects have the greatest impact on the system performance measures. This frequently helps identify opportunities to further develop some areas for better accuracies or to simplify others in order to reduce model execution time.

The numbers of actual simulation runs, initial conditions and number of variables are all selected in advance, based on statistical analyses. The results of the simulation runs are then interpreted and documented. As in any area connected with T & E, the documentation is especially important, because those who immediately use its results need to understand those results in the context of its limitations and assumptions built into the M & S effort.

## 6.12   COMPUTER AIDED T & E TOOLS

Computer aided tools which have been proven to be very useful in the management of T & E programs are becoming available. Functions for which tools are available include:

- requirements analyses
- functional flow-down of system level performance requirements
- specification trees
- risk analyses
- correlation of performance requirements to T & E events.

These tools are basically application databases with one or more structured analysis and design methodologies, suitable for use in systems engineering. Like any computer based tools, costs versus benefits, tailoring to the particular applications, adequacy of host equipment, training and continued availability of qualified personnel, and anticipated extent of usage, are all to be considered before investing in any tools (Smith 1995).

## 6.13 SUMMARY

Even as testing itself is becoming recognized as a systems engineering discipline, there are already specializations of expertise within T & E, based on different techniques of T & E. As discussed in this chapter, these areas of specialization have different levels of experience.

## REFERENCES

Beizer, B. (1983) *Software Testing Techniques*, Van Nostrand Reinhold Company, New York.

Fath, J. L., Mann, T. L. and Holzman, T. G. (1994) A Practical Guide to Using Software Useability Labs: Lessons Learned at IBM. *Behavior and Information Technology*, **13**, pp. 94–105.

Hammer, W. (1980) *Product Safety Management and Engineering*. Prentice-Hall, Englewood Cliffs, N.J.

Holden, M. and Girardi, L. F. (1991) Wind Tunnel Tests Help Take the Drag Out of Skiing. *ITEA Journal*, **XII**, (1). pp. 32–33

Law, A. M., and McComas, M. G. (1990) Secrets of Successful Simulation Studies. *Industrial Engineering*, May.

Meth, M. A. (1994) Practical 'Rules' For Reliability Test Programs. *The ITEA Journal of Test and Evaluation*, December1993/January 1994 issue.

Pressman, R. S. (1987) *Software Engineering: A Practitioner's Approach*, 2nd edn, McGraw-Hill Book Company, New York.

Smith, J. M. (1995) *System Engineering Management Tools*, University of South Australia, Australian Centre for Test and Evaluation, August.

*USA Today*, June 14, 1995.

US Code (1986) Title 10 US Code Section 2399.

USDoD (1985) *Testability Program for Electronic Systems and Equipments*. US Department of Defense Military Standard 2165, January 26.

USDoD (1995) *Software Development and Documentation*, US Department of Defense Military Standard 498, August.

Wolkomir, R. (1995) Sitting on Our Stead: Crash Dummies Take the Hard Knocks for All of Us. *Smithsonian Magazine*, July.

# 7

# The Seven Best Practices

*If the only tool you have is a hammer, you tend to see every problem as a nail*

Abraham Maslow

Since the T & E program is to be used to confirm proper system performance, it must have both integrity and credibility. It must be accepted by all involved as being capable of accurately comparing the system – and the development program – to 'ground truth'. People often confuse the success of the T & E program with how favorable the T & E results are. The former is a function of factors such as how well the test was planned, how valid the test methods were, how rigidly the test conductors adhered to the test procedures, and how rigorous the data analysis and reporting were. The latter is a function of the design and engineering of the system. It is possible to have an excellent T & E program – one that accurately measures the system's performance – and yet have a development program that fails, i.e., it falls to provide a system that meets the minimum acceptable performance requirements. Today, this distinction may be of interest to very few. A T & E professional taking credit for a great T & E program in spite of poor system performance is analogous to a surgeon saying the operation was a 'success' even though the patient died. In this chapter, we will discuss some of the 'best practices' gleaned from the conduct of T & E programs that were widely accepted as successful.

## 7.1 RECENT EXPERIENCE

As described in Chapter 1, we have only several decades of experience conducting the type of complex system T & E programs which are the focus of this book. And even within that time frame, the overall philosophy of T & E for these programs has changed. Those changes for the most part have been driven not by the experiences of the recent past, but by the need to keep pace with ever advancing technologies. The complex systems of the 1960s and 1970s were well suited to the emphasis placed on system-level integration testing and the new commitment to conducting full operational T & E. Those systems were definable, and their T & E programs could be meaningfully and readily bounded. But today, we are developing systems that cannot be tested – at least, not fully tested.

As we look at today's generation of successful T & E programs and try to capture what made them work well, we find that the salient features are not easily described – at least in terms such that a new program can readily adopt them as a road map. Putting hardware through the full spectrum of environmental tests, putting computer software through stress testing and regression testing, and putting the entire system (with the users participating) through full operational T & E no longer – by themselves – can provide confidence in the development program. T & E can no longer be considered primarily a catalog of T & E methodologies and procedures that are judiciously mixed and tailored to a program's risks. Instead, T & E must be viewed first and foremost as a *process*. More is needed than doing early, detailed planning. A look at the lesson's learned from today's successful T & E programs shows they have the following characteristics.

## 1

System requirements are well defined and well understood, the associated risks are generally agreed upon, and the linkage between the risks and the T & E requirements is kept current. One could argue that involvement in setting system requirements is outside the scope of the T & E program – i.e., the T & E program planning begins where the requirements definition leaves off. That can no longer be the case, and any program that subscribes to that philosophy is bound to have problems in T & E. Surveys of major development programs in the last decade show that usually half of the serious deficiencies that are not uncovered or corrected by the end of the program are attributable to problems or issues in defining system requirements.

## 2

Comprehensive planning for T & E is started early. Such planning includes allowing for the contingencies that will be necessary when – as is inevitable – things go wrong. In any type of program, some of the risks can be anticipated (the 'known unknowns'), and some cannot (the 'unknown unknowns'). Whether planning an extended 10-year T & E program, or planning a two-week test event, allowance must be made for properly identifying technical problems, correcting them, and validating the corrections. There are always pressures for the program team to underestimate the time and resources necessary for testing. Most program managers are, by nature, success oriented. The T & E people must be in a position to counter these pressures with knowledgeable and reasonable estimates of what it will take to do the job right. Such estimating should be based primarily on engineering considerations, rather than being unduly influenced by schedule commitments. Even when schedules and budgets allow for contingencies, detailed plans to make use of those contingencies must already be in place when the problems occur. Otherwise, large resources can be ineffectively spent if a logical plan is not being followed to get the effort in hand back on track. Since the start of some tests may depend on the completion of others, the tests most critical to the overall program schedule should be identified, and work-around plans should be developed for use in case of unanticipated problems. For example, work-arounds may involve changing event sequences, applying more test assets or using additional test facilities. Of course, contingency plans must be continually updated and refined as the program progresses to ensure both that they reflect engineering and management experience being gained and that they remain properly tailored to the ever-changing risk assessments. Also, planning for all T & E must start early. Usage profiles, for example, must be developed and analyzed early to be available for planning reliability testing, design-limit testing and life testing that will be conducted through the program, at the system, subsystem, and component levels. In addition to early planning for testing during design and development (and the continual refinement of those plans), planning for production and in-service testing must also begin early, i.e., early in design. Chapter 5 discussed a document called a T & E Master Plan, the preparation of which helps focus planning at the program level and provides a point of departure for lower level T & E plans.

**3**

There is a strong company-wide and program-wide commitment to using the T & E program to identify problems as early as possible – even at the expense of some design and engineering work if necessary. A building-block approach of 'Build-a-little/Test-a-little' must be embraced throughout the organization, and at all levels there must be a sincere interest in letting the T & E results pace the program. There are wide differences in the roles to which T & E is relegated in today's development programs. All programs advertise that T & E is being used to achieve the typical objectives, but the reality is that many, if not most, organizations let schedule and cost considerations unduly pressure and constrain the T & E program and the ability to adequately investigate and incorporate the recommendations that come from the T & E program. The following example – perhaps an extreme one – makes this point (Romer 1994).

### Case example: Republic of China Indigenous Defense Fighter

In 1984, the Republic of China (ROC) (Taiwan) government decided to develop an Indigenous Defense Fighter (IDF) for its Air Force. The original program called for the construction of four prototype test aircraft, three single-seat and one two-seat, for the conduct of joint Developmental T & E and Operational T & E. This was one of several new indigenous programs, for which the ROC adopted many of the features of the US Defense Department's acquisition process, including the conduct of Operational T & E by a government organization separate from the development organization. Instead of the traditional five phase program, the ROC used only two: full-scale development and full-scale production. This short cut was based on their assessment that development risk was low and that the subsystems would arrive ready for installation. This approach, however, incorrectly minimized the need for system integration. In addition, responsibility for the full-scale development phase of each subsystem was assigned to one organization and the responsibility for production of everything but the missiles was the assigned to another.

In the Chinese culture, the psychological loss of face is the worst of all possible outcomes in both personal and business relationships. Losing face is to be avoided at all costs. And, for one person to cause another to lose face would invite retribution at some late time. As a consequence, perfection is expected the first time anything is done, and one is reluctant to point out a lack of perfection. Taking risks is avoided. Decisions are delayed as long as possible to avoid risks and the accountability for risks. In China, testing is conducted to validate the perfection of the design – not to uncover problems. During stability and control testing on the IDF, the program manager was reluctant to authorize a high angle of attack flight – lest the aircraft enter a spin requiring a deployment of the spin recovery parachute. This would have been a very public admission that the aircraft design wasn't perfect.

During a test flight in July, 1991, one of the prototype IDF test aircraft crashed. Although there appeared to be plenty of opportunity to conduct a thorough investigation, one was not done. The official cause of the accident was listed as the unusual wind conditions that day, at that time, and at that altitude. No design or manufacturing process flaws were identified or corrected. Of course, there is now a hole in the flight envelope and no one deliberately flies at that altitude and airspeed.

In the Chinese culture, information is power and protecting it is paramount. Although the technology is available for computer networks, they do not exist or are rarely used. Everyone keeps a personal record. When a problem arises, rather than turn to records or data banks, a new collection system or analysis process is created. A credible failure reporting and corrective action process did not exist for the IDF program. In 1992, the program was cut from a planned 240 aircraft to 120. Funds and talent that may have eventually improved the aircraft were diverted to other programs.

For T & E during a development program to be fully effective, all of the participating organizations need to adopt a positive attitude towards the T & E. First, they need to understand and advocate the conduct of good T & E as early as possible to identify design weaknesses when there is maximum flexibility to correct them. Secondly they need to appreciate the 'checks and balance' benefits a rigorous and visible T & E program brings to the systems engineering efforts. The T & E program should not be viewed as an obstacle, but as a tool. Thirdly, any T & E people that are organizationally separate from the development team need to be kept apprised of design problems and solutions, as well as trade-off decisions, so they can be judicious in their failure analysis and in their recommendations for correction.

---

### Case example: BOEING 777 aircraft program

In the mid 1990s, Boeing designed the 777 passenger aircraft (Figure 7.1), essentially the replacement for the 747. Initially, when Boeing was setting up the program, it did an extensive survey of its 747 airline customers, asking what they would most like done differently in the 777 program. They were surprised that by far the strongest recommendation was that the aircraft be much more rigorously tested before being placed in service to avoid the continuous design problems uncovered immediately after the 747 was placed in service. Mr Gordon McKinsey, United Airlines program manager for the introduction of the Boeing 777, said, 'There's a famous quote in [the airline] industry . . . ''The 747-400 is a beautiful airplane. I can't wait until Boeing finishes building it.'' ' (Dornheim 1994)

For the 777, Boeing conducted almost three times the flight test hours it performed on the 767. The majority of this testing was 'service ready' testing not required for FAA certification, but conducted to identify design problems before the aircraft were delivered. By investing more in this pre-delivery testing, Boeing saved much money – and easily recouped that investment – by not having to fix those problems on up to 50 already completed aircraft. Aside from the service ready tests, the FAA required 90 flights on the lead customer's routes to gain Extended Twin-Engine Operations (ETOPS) approval. These were non-revenue operational T & E flights, piloted by United Airlines crews and maintained by United mechanics, but paid for by Boeing.

Another investment in T & E for Boeing has been the construction of an Integrated Aircraft Systems Laboratory in Seattle, Washington (see section 4.3.2). Boeing credits the lab with the 777's certification by the Federal Aviation Administration for ETOPS upon its first delivery to United Airlines in 1995. Boeing also believes that the lab will give them a significant price advantage against Airbus and McDonnell Douglas by helping to minimize expensive production floor changes and cut warranty costs. During advanced testing, multiple systems are often integrated and 'flown' much like a real aircraft from one of at least 11 flight simulators.

In some cases, this favorable mentality toward T & E does not come naturally – at least initially. In computer software and hardware being developed for home use, a few months advantage in delivering a new program or capability to the market-place can make a huge difference in the overall sales. Advances and improvements are coming at such a rapid pace that there is great pressure within the developing companies to get the products 'out the door,' even when it is known that there are some errors in them. In 1994, after much advanced publicity, IBM released the Pentium model of its Personal Computer. Within six months, a much publicized defect was found in the Intel processing chip that made arithmetic errors in some complicated calculations. Intel eventually promised to replace the defective chips in all computers that had it, free of charge.

## 4

The T & E network is well coordinated. The roles and responsibilities of each participant are well defined, yet all are willing to work as a team for the good of achieving a successful product. The program manager, or an agent who is clearly empowered to make decisions for the program manager, holds regular T & E progress reviews. Status information is readily and regularly exchanged, greatly enabled by electronic mail and teleconferencing. Staffing decisions and tasking correspond to the varying needs of the program. Costs and schedules are closely monitored by all, they are modeled where possible, and variations are reviewed and well understood. Entry and exit criteria for T & E events are well known. For major events, Test Readiness Reviews are held and key program personnel participate. There is an openness to identifying and discussing problems and to considering all solutions.

**Figure 7.1**
Boeing 777 aircraft

Major suppliers and subcontractors are treated as fully fledged members of the development team. Also the financial sponsors, and typical system operators and product customers, participate in planning the T & E program. They have compatible reporting mechanisms, evaluation metrics, incentives, and commitments to quality as does the rest of the program team.

## 5

T & E resources are closely managed so they don't fall on the critical path. Such resources include: test articles properly configured, spare parts and system test equipment (if appropriate for the test), computer software (including that of the system, of external stimulators, and of data analysis), system operators and maintainers, system documentation, verified test procedures (including data sheets), analysis methodologies and tools, adequately validated models and simulations, test facilities, test sites, calibrated instrumentation, and qualified test directors.

## 6

There are good policies and procedures for ensuring the development of quality computer software. The overall quality assurance program is process oriented, rather than product oriented, with well designed, tailored metrics. An independently trained quality assurance staff is organizationally separate from the software developers. There is a strong emphasis on and incentives for finding and fixing software problems at the earliest time and at the lowest level of development. Peer reviews are conducted on all major segments of the program. The process for the delivery of the program code has well defined procedures. And there are well accepted, automated, configuration control procedures.

## 7

Modeling and simulation (M & S) are used, wisely. M & S data are visibly kept distinct and separate from T & E results using actual system hardware and software. However, the M & S must be used in conjunction with T & E throughout the program to help optimize the effectiveness of the test events and test procedures as they are planned, as well as to extrapolate the T & E results into regions of the performance envelope that are not being tested. The increasing dependence on M & S and the mixing of actual T & E data with M & S data in the later stages of a development program make it critical that everyone participating in the program be aware of and maintain the distinctions between the two. In US Defense Department programs, Public Law specifically prohibits the use of M & S results as a substitute for operational T & E (US Code 1987). To lose track of the distinction of M & S results from actual T & E results, as well as not to fully consider the precision and fidelity of the particular models in interpreting the results and drawing conclusions from them, can result in pitfalls such as the following:

(1)   unknowingly venturing outside the area of the model that has been validated, introducing unplanned risk and uncertainty

(2)   relying on heart-of-the-envelope performance data

(3)   using specification values instead of performance data

(4)   continually making worst-case assumptions in an effort to be conservative

(5)   assuming independence between events that actually are not independent.

It is important that the development team, and to some extent the financial sponsors of the program and the prospective users of the system, understand the credibility of the M & S efforts planned. In evaluating this credibility, it is vital to understand the purposes for which the model was originally built and its strengths and weaknesses for this application to T & E – if it was not built for this specific T & E program. This involves researching what the assumptions of the original model developer were. It is also important to understand the credibility of the data to be used in the model: what was their source? who certified them? The simulation runs in themselves must also be credible: does the customer agree that they that are meaningful and realistic? There should be a deliberate process whereby the customer certifies that agreement. To what extent do the results from M & S agree with data from the 'real world' or the results of other models that have an accepted level of credibility? What is the statistical significance of any 'real world' data used in the M & S, and the statistical significance of the M & S data itself? If multiple models are used, what is the credibility of their linkages?

## 7.2   SUMMARY

Compared with many areas of industry, some of which have well progressed over the last century from being a theoretical art to being a craft, T & E is still in its infancy. Its growth and development are both pushed and pulled by the disciplines that it supports – many of which themselves are just evolving. For a long time to come, the success of planning and executing a T & E program will be critically dependent on sound professional judgement, which in turn must be based in large part on using approaches that have proven to work in the past. Identifying and reapplying 'best practices' such as the seven listed here will continue an important part of the advancement of T & E.

# REFERENCES

Dornheim, M. A. (1994) Service Ready Goal Demands More Tests. *Aviation Week and Space Technology*, April 11, 1994. p 44

Romer, R. A. (1994) C.P.L., Operational T & E in the Republic of China – A Clash of Cultures, paper presented at the *Annual Symposium of the International Test and Evaluation Association* in Baltimore, MD, October 4.

US Code (1987) Title 10 US Code, Section 2399.

# 8

# Summary: the Future of T & E

*The trouble with the future is that is usually arrives before we are ready for it*

Arnold H Glasgow

The changes in T & E have been so dramatic in the last few decades, they prompt the question, 'Where is it going in the future?' This chapter discusses that question in four over-arching dimensions in which it can be expected to expand and mature: technology, tools, image and professionalism.

## 8.1 TECHNOLOGY

Dr Malcolm Currie, chairman emeritus of Hughes Aircraft Company, has offered the following predictions for the year 2025 (Berry 1993):

- air traffic control by satellites,

- free trade worldwide,

- push button warfare at long distance that will make battle casualties practically unacceptable,

- computers a hundred thousand times more powerful than those of the 1990s,

- real-time universal language translation for both voice and text, which will reduce cultural differences,

- eradication of most diseases in the world through computer power simulating biological systems and synthesizing new molecules,

- accurate global and local weather forecasts up to one month in advance,

- virtual reality (advanced computer based simulations of the environment)

progressed to the point where consumers can buy disks with three-dimensional databases that will allow them to experience vacations and cultural events while sitting at home, and

- real-time person-to-person communications from any point in the world to any other point.

As we dare imagine the systems and products that technology will bring us in the years ahead, we must also imagine the growing consequences of failures in them. If the last few decades are any indication of what lies ahead, consumers, industry and governments will insist on disciplined, verifiable, T & E to lessen the risks of failure.

## 8.2   T & E TECHNIQUES AND TOOLS

It is technology that will drive T & E to new frontiers, and it is technology that will enable T & E to be effective at those frontiers. The very technologies that will drive the need for more advanced T & E methodologies will also enable those advances. Continued advances in computer size, memory, processing power and speed seem to be at the core of nearly all of our predictions. As it has in the past, T & E of computer based systems will have to keep pace with the increasingly complex design of those systems. We can expect computer technology to continue to grow rapidly, challenging us to find testing techniques to keep up with new design techniques. One example of a new advancement is that of virtual reality, the computer depiction of environments – frequently multi-sensory, such as animation on a display coupled with sound or touch through helmets and gloves – and the capability of interaction of the user with that environment. The applications of virtual reality are becoming more common, and have the potential eventually to improve system and product designs as well as to enhance job performance and training.

Another computer related technology area that will rapidly expand is modeling and simulation. One new dimension into which it has already moved is that of Advanced Distributed Simulation (ADS). The continuing evolution of computer capabilities provides for far more accurate, real-time simulations. It is now possible to have different simulations at different geographical locations acting together. High-speed communication networks, common network interface translation devices, and emerging standard data protocols are key to this interoperability. This has been useful particularly in military system developments, where it has allowed the creation of realistic environments in which computer simulations, real equipment and systems, and humans work together in synthetic battlefields for training, system development and T & E (Sikora and Coose 1995). Those simulations in which real people operate real systems in an environment that is at least partially simulated are called 'live' simulations. Those simulations in which real people operate simulated systems are called 'virtual' simulations. Those in which simulated people operate simulated systems are called 'constructive' simulations.

In the hardware arena, there are now ways of testing electronic circuit cards that can reduce some of the time, cost and error associated with completely human testing. One of the most recent techniques is called modal testing. Using temporal logic – a language for describing a system's behavior over time – modal checking finds circuit errors that escape the simulation tools that chip makers have traditionally used, according to the Carnegie Mellon University researchers who have developed it. This approach lets testers thoroughly examine chips of enormous complexity that can produce many different states, as many as 10 to the 10th power (Hendricks 1995).

T & E has become a more recognized and integral part of systems engineering in complex systems, but 'how much testing is enough' will always remain a difficult judgment call. Better design tools, in particular computer aided engineering, greater use of modeling and simulation, and more efficient use of test facilities will all tend to reduce the ever increasing costs of some T & E. But such reductions will frequently be offset by the growing need for new, higher levels of interoperability testing as systems continue to become more interactive and interdependent. And none of our new tools and techniques will ever obviate the need for the attention to detail required at every dimension of systems engineering. The root cause of the famous Intel chip problem (Section 7.1) was that an engineer copying a type of 'look-up' table on a computer wrote a script in C language to download it into a Programmable Logic Array (PLA). Due to an error in the script, 5 of the 1066 entries were not downloaded. Nobody checked the PLA to verify that the table had been copied correctly (Halfhill 1995)!

## 8.3   RESEARCH IN T & E

With the role of T & E for advanced technologies in both test items and test methodologies, the T & E community has recognized that it must marshall and share its knowledge of any research in T & E that could have wide application. For example, the substantial investments needed because of facilities that are rapidly becoming outdated can be partially offset by the new methodologies in simulation and internetting of older and newer facilities. The strengths of the capabilities of some facilities can be leveraged for others. A number of programs have been initiated to address this.

The University of South Australia has established the Australian Centre for Test and Evaluation as a technology center in order to provide a focus for T & E in the Asia-Pacific region. The aim of the Centre is to develop the professionalism and skill level of T & E practitioners by high-level education and training, technology transfer, research and development, and consulting and project management.

In 1995, Georgia Tech Research Institute (GTRI) in Atlanta established the T & E Research and Education Center (TEREC) to 'advance knowledge in T & E and distribute it through education'. The center fosters Georgia Tech activities that track technological trends in T & E, carries out specialized research and development activities, analyzes the benefits of test capabilities, and conducts workshops and short courses in T & E subjects.

## 8.4    THE IMAGE OF T & E

While the need for more formalized T & E will become more recognized and accepted, T & E will remain a function that must continually be promoted to managers not only when a system development is initially planned, but throughout the program. Government and industry performance requirements as well as system interface and interoperability standards will continue to grow, and along with them the requirement to verify compliance through T & E. This growth will nurture the attitude that since many T & E requirements transcend the goals of the individual program, the T & E professionals cannot truly be part of the inner core of the development team. Even for the T & E requirements that are part of the program's systems engineering, the T & E managers will continually need to 'sell' the level of T & E they think is necessary and sufficient.

A commercial system developer has to strike an intelligent balance between the *cost of perfection* and the *cost of correction*. In the 1970s, US car makers were not striking that right balance, and consumers were deserting in droves to Japanese makers. Now US auto makers are doing better. So too with computer software. Some say the more serious mistake made by Intel with their processing chip introduced in Pentium computers in 1995 was not the undiscovered software floating point error (few users even know what it means), but with their marketing error in hesitating before offering no-questions-asked replacements. Even though the risks are higher today than ever before for introducing systems and products of inferior quality, the stakes are higher than ever before for getting products to the market-place quickly. Being first may well give a developer an insurmountable edge, particularly in high-volume, low-margin products like software for personal computers. Such time-to-market pressures promote constantly changing product requirements and actually create a market for yearly program updates. The updates themselves then become sales opportunities: opportunities to *sell* corrections to previous program errors. Such opportunities can detract from the objective of ensuring high quality in the original versions of new products in the first place.

Nevertheless, while T & E will be a separate function in the development processes, it will nonetheless be integral to getting the system to the customer. The most dominant change in the management techniques and styles in development programs that have emerged in the last few decades is a greater dependence on teaming. 'Integrated Product Teams' (IPTs), composed of the program manager, sales department, quality assurance and T & E, and the marketing department work closely together to reduce costs and get satisfactory products to the customer as quickly as possible. At the corporate level, 'integrated *process* teams' work together to maximize the efficiency and effectiveness of every process and procedure. And *between* corporations, there is teaming too for the overall good of the each member. The teaming of Boeing, General Electric, United Airlines and the Federal Aviation Administration is a prime example. T & E will be a very evident part of the teaming that takes place at every level of system development.

## 8.5   T & E AS A PROFESSION

As T & E increases in formality and discipline, and as the inevitable highly visible problems that might have been prevented with better T & E are uncovered, T & E will become more of a profession. And the T & E professionals will continue to network themselves. Evidence of this is already apparent in the remarkable growth of associations such as the International Test and Evaluation Association (ITEA), headquartered in Fairfax, Virginia. Founded in 1980, ITEA now has chapters as well as corporate and individual members worldwide, with regular symposia and workshops, a renowned quarterly publication, and a prestigious awards program. ITEA and professional associations like it will be a catalyst for a global exchange of information on T & E related subjects. More information about ITEA is contained in Appendix D.

Courses in T & E are becoming common. Many of the early courses were developed by the US Defense Department in response to the initial establishment of the Department's T & E policies in the 1970s. Among the earliest were courses conducted by the Defense Systems Management College, in Ft Belvoir, Virginia. ITEA initially started conducting short (one-day or less) courses in conjunction with its symposia and workshops and soon expanded them into a separate education program consisting of several three- and four-day courses per year. In the mid-1990s, Georgia Institute of Technology in Atlanta established a program to provide a subspecialty in T & E as part of its programs for the Master of Science in Systems Analysis (in its School of Industrial and Systems Engineering) and a Master of Science in Electrical Engineering (within its School of Electrical Engineering). Taking a number of required and elective courses in T & E allows the student to receive a Certificate in T & E. The University of South Australia has begun a cooperative program with the Georgia Institute of Technology to offer postgraduate courses in T & E.

## 8.6   CONCLUSION

As we said in the Preface, we are now building systems and products that we cannot fully test. Nonetheless, we still need to be able to fully evaluate them.

Without diminishing the passion for detailed, disciplined work that will always be at the heart of successful systems engineering, we need to embrace the expanding role of T & E and its ever increasing interrelationship with computer aided engineering, modeling and simulation, and whatever technology offers us in the future – to ensure safe, quality systems and products into the 21st century.

## REFERENCES

Berry, F. C., Jr (1993) *Inventing the Future*. Bassey's (US), Washington. p. 1

Sikora, J. and Coose, P. (1995) What in the World is ADS? *Phalanx* (Bulletin of the Military Operations Research Society), **28**, (2).

Hendricks, M. (1995) Pentium Post-mortem: Checking the Chips. *Popular Science*, April. p. 49

Halfhill, T. (1995) The Truth Behind the Pentium Bug. *Byte*, March. pp. 163–164

# Appendix A

## Sources of T & E Specifications and Standards

The following is a listing of major organizations that have published T & E specifications and standards and have made them available for general use.

American Society of Testing and Materials
(ASTM)
1916 Race Street
Philadelphia PA 19103
USA

American Society of Mechanical
Engineers (ASME)
345 East 47th Street
New York NY 10017
USA

American Society for Quality Control
(ASQC)
611 East Wisconsin Avenue
Milwaukee WI 53202
USA

British Defence Standards
Ministry of Defence
Stan 1 Kentigern House
65 Brown Street
Glasgow
Scotland G2 8EX

British Standards Institution (BSI)
389 Chiswick High Road
London
England W4 4AL

Chinese National Standards
3rd Floor 185 Section 2
Hsin Hai Rd
Taipei, Taiwan

Canadian Standards Association
178 Rexdale Blvd
Etobicoke
Ontario M9W 1R3
Canada

Electronic Industry Association
2500 Wilson Blvd
Arlington VA 22201
USA

European Committee for Electro-
Technical Standardization
(CENELEC)
Rue de Stassart 35
Bruxelles, B-1050
Belgium

European Committee for Standardization
(CEN)
Rue de Stassart 36
Bruxelles, B-1050
Belgium

European Organization for Civil Aviation
  Electronics
11 Rue Hamelin
Paris
France

European Telecommunications Standards
  Institute
Route des Lucioles
06921 Sophia Antipolis
Cedex
Valbonne 06921
France

International Electrotechnical
  Commission
3, Rue de Varembe
PO Box 131
Geneva, CH-1211
Switzerland

Institute for Interconnecting and
  Packaging Electronic Circuits
7380 North Lincoln Avenue
Lincolnwood IL 60646
USA

Institute of Electrical & Electronics
  Engineers
445 Hoes Lane
Piscataway NJ 08855
USA

International Electrical Testing
  Association
PO Box 687
Morrison CO 80465
USA

International Organization for
  Standardization
Case Postale 56
Geneve 20, CH-1211
Switzerland

International Telegraph & Telephone
  Consultative Committee
Place des Nations
Rue de Varembe
Geneva, CH-1211
Switzerland

Japanese Industrial Standards
1-24, Akaska 4
Minato-KU
Tokyo, 107
Japan

Joint Technical Committee: ISO/IEC
  JTC1
Case Portale 56
Geneve 20, CH-1211
Switzerland

Korean Standards Association
13-31 Yoido-Dong
Youngdungpo-gu
Seoul, 150-010
Republic of Korea

NACE International
1440 South Creek Drive
PO Box 218340
Houston TX 77218
USA

National Environmental Balancing Bureau
8575 Grovemont Circle
Gaithersburg MD 20877
USA

National Electrical Manufacturers
  Association
1300 North 17th Street
Suite 1847
Rosslyn VA 22209
USA

National Fire Protection Association
One Batterymarch Park
PO Box 9101
Quincy MA 02269
USA

North Atlantic Treaty Organization
Ministry of Defence
Kentigern House
65 Brown Street
Glasgow
Scotland G2 8EX

Robotics Industries Association
PO Box 3724
Ann Arbor, MI 48106
USA

Society of Automotive Engineers (SAE)
    International
400 Commonwealth Drive
Warrendale PA 15096-0001
USA

Saudi Arabian Standards Organization
PO Box 3437
Riyadh 11471
Saudi Arabia

Society of British Aerospace Companies
29 King Street
St. James's
London
England SW1Y 6RD

Standards Association of Australia
PO Box 1055
Strathfield NSW 2135
Australia

Standards New Zealand
155 The Terrace
Wellington 6020
New Zealand

System Safety Society
5 Export Drive, Suite A
Sterling VA 22170-4421
USA

# Appendix B

## Qualification Requirements for T & E Personnel

Listed below are the usual qualifications, i.e., knowledge, skills and abilities, typically needed by the T & E professional today for managing and executing T & E as part of the development program for a complex system or product.

### General

Knowledge of and experience in the management of development programs.

Knowledge of financial management techniques in development programs.

Ability to translate system requirements into test requirements, and to optimize the articulation of system requirements in terms best suited for T & E.

Skill and experience in cost estimating T & E events.

Ability to structure a T & E strategy that satisfies the customers and users.

Proven ability to garner and sustain the credibility of all concerned in the integrity of T & E and the validity of its results.

Experience in planning, executing and reporting T & E events.

Knowledge of the latest computer software development metrics and management tools.

Knowledge of the use of the latest analytical tools for use in T & E planning.

Ability to identify the sources of risks in a development program well in advance and to determine how the T & E program can best mitigate some or all of those risks.

Experience in effectively using modeling and simulations.

Knowledge of the techniques used to optimize testability in system designs.

## Program-Specific

Understanding of the need or opportunity that generated the system performance requirements.

Knowledge of existing T & E standards and procedures that have a bearing on the program.

Knowledge of any statutory requirements that could impact the T & E program.

Knowledge of and experience with modeling and simulation projects.

Proven ability to capture, integrate, and maintain a current understanding of the financial sponsors', the suppliers', the customers' and the users' needs and desires.

Knowledge of existing test facilities, including their capabilities and limitations, that may be used in the program.

Knowledge of independent laboratories that may be useful.

Knowledge of the latest technology in data collection and data analysis needed by the program.

# Appendix C
## Case Example of a T & E Master Plan

### C 1   INTRODUCTION

As described in Chapter 5, a T & E Master Plan (TEMP) is a necessary top-level document that serves as the ultimate reference and point of departure for lower-level T & E program planning. It is the bridge between the system performance requirements and the verification that those requirements are being met, as well as an important written agreement between the heads of all organizations that are major participants in the program. For a development program for a complex system or product, it is a 'must.'

Based on experience in developing TEMPs, the following guidelines have evolved:

(1)   A TEMP should be kept relatively short. If it is to be an *executive-level* document that reflects the agreements of and is visibly binding on the heads of the major organizations involved, it should be no more than 25 pages. The US Department of Defense guidelines specify no more than 30 pages, not counting appendices. However, TEMPs that do exceed 25 pages could readily be reduced to that number if they were stripped of detailed material that is informational only (i.e., it is not part of the major agreements between the participants), but instead has been inserted at the insistence of staff people in the organizational hierarchy. Before page limits were put on TEMPs in the US Defense Department, some exceeded 100 pages, which discouraged executives from reading the document. Since it was assumed by most people that the executives probably did not read it, it came to be viewed as an agreement between staff people, greatly limiting the perception of a commitment by upper management. On the other hand, if the document is agreed to and signed by the key executives, it 'sends a message' to everyone about the commitment of the leadership of the key organizations to having a truly effective T & E program.

(2)   There is a need to and value in conscientiously differentiating between the top-level T & E planning that needs to be captured in the TEMP and the lower-level planning that should *not* be in it. If the planning of these levels overlaps and is not kept distinct, each will become outdated more quickly – undermining their value in helping to coordinate and guide the work of the overall team. Overlaps will also lead to time-consuming confusion and perhaps disagreements regarding what changes will require a (necessarily burdensome) renegotiation of the TEMP. Unless the TEMP is limited to the top-level planning, it cannot readily serve its role as the guiding document that gives focus and direction to the myriad of lower-level planning work.

(3)   The most important part of the TEMP is the description of the criteria for evaluating system performance as acceptable or not. As described in Section 4.1, the pyramid of criteria need to be *engineered*. Only the top-level (pass/fail) criteria should be included in the TEMP since it is a top-level contract between major program participants. If second-tier or dependent parameters and criteria are put into the TEMP, the achievement of the performance they represent will inevitably be misunderstood by some to be equally as important as the first-tier criteria, thereby skewing the evaluation results. This can easily happen in a large program organization with many participants. Another reason for carefully limiting the pass/fail criteria is because the design and engineering organizations need the flexibility to be able to adjust the lower-level performance criteria as they make design trade-offs, without negotiating with the executive signatories to the TEMP. Criteria should be quantified wherever possible so as to enable the ready accomplishment of the analyses and evaluations. Also, the criteria should be testable. If some part of the performance envelope is not testable because of limitations (e.g., in instrumentation, cost, time, security, etc.), the criteria should reflect those limitations. Any special conditions, specific system configurations, or test methods that uniquely pertain to a particular measure and value should be clear in the TEMP or a referenced document (e.g., a T & E standard) so as to reduce the possibility of misunderstandings later when the results are being interpreted. For computer software-intensive systems, metrics to be used to track the progress during development should be described since full operational performance of the software cannot be meaningfully demonstrated until the end of the development effort when it is installed in a full-up production-representative version of the system.

(4)   The objectives (e.g., questions to be resolved) of each major test event, the configuration of the system to undergo T & E, the planned scope and location of the T & E, and limitations to the scope that are known in

advance should all be described in a narrative form. The planned use of models and simulations to supplement actual T & E should be described as well as the extent to which each model and simulation will be validated, and how it will be validated.

(5)  Special T & E resources, the availability of which are critical to the accomplishment of the objectives of the major T & E events, should be listed, along with the plans for obtaining them or getting ready access to them when they are needed.

(6)  The TEMP needs to be updated throughout the program. Its contents will differ significantly from one stage of system development to the next.

## C2  – A TEMP

In this Appendix, we have provided a TEMP for a fictitious program – a new US battery-powered automobile whose sales will hopefully bring electric vehicles into significantly common usage. It is the first version of the TEMP, reflecting what the planning might be at the beginning of the program.

A typical TEMP should be between 20 and 25 pages. This example is much shorter than that, so it can be as readable as possible. If this were a real program, a much greater depth of discussion would be necessary. We have limited our discussion in order to illustrate the points about good T & E planning and plan writing that we are making here.

At the end of this TEMP is a discussion of how it demonstrates many of the features of a good Master Plan and of how the T & E it describes demonstrates many of the characteristics of a good program.

TEST AND EVALUATION MASTER PLAN FOR THE STIMULUS AUTOMOBILE

First Version: Program Inception

Prepared by the T & E Manager:
Approved by the Program Manager:
Approved by the Quality Assurance Manager:
Approved by the Design Manager:
Approved by the Production Manager:
Approved by the Company President:

## C2.1   Description

*The Program*

In the year 2005, four US automobile manufacturers formed a joint venture company to design, manufacture, sell and service a significantly new, quality automobile with dramatic improvements over prior models. Sales of US automobiles had been lagging and it has been announced that the European Union (EU) plans to design a new automobile for worldwide marketing. This latter automobile has been planned as the first cooperative development program among the member nations of the EU. There is conjecture that the EU nations are committed to secretly heavily subsidizing the development to ensure its success. The US auto makers were concerned about this competition, and therefore formed this joint-venture company.

Several years earlier, the US Government stepped up research and provided incentives for more widespread use of electric vehicles (EVs). Several significant technical breakthroughs in battery technology occurred greatly increasing the chances of successfully mass marketing automobiles with EVs. With the help of federal grants, states started providing charging facilities for gas stations, public parking facilities and private homes; and providing preferred parking, car-pool credits, and tax incentives for users of EVs.

The joint venture company decided that the major design improvement around which their new auto would be designed and marketed would be electric (battery) power. It obtained a substantial grant from the US Department of Energy. The schedule objective is to introduce the automobile at the 2010 World's Fair in Sydney, Australia. This document is the original TEMP, prepared shortly after program initiation. Therefore many of the technical details remain to be decided.

*Summary Analysis of Risks*

The major risks in the program relate to the development of the electric power system. The most significant risks are, in descending order:

- risk of not meeting stated performance requirements. Work on regenerative braking, electronic controls, heaters, and air-conditioners is also needed so their power demands do not interfere with the car meeting the minimum ranges between recharges.

- risks that meeting stated performance requirements will still not capture significant share of the market. Availability of sufficient number of recharging facilities could hamper sales.

- risk that the scheduled introduction at the 2010 World's Fair will not be met. Missing this highly visible objective could reduce consumer credibility and interest, as well as detracting from the success of the aggressive advertising campaign.

- risk of cost overruns. Full extent of US Department of Energy subsidies will not be known at the outset.

## The Requirements

The automobile has been named the 'STIMULUS'. It is intended to be a sporty looking passenger automobile, attractive as a family car. There will also be a left-hand drive version for export outside the United States.

Specific requirements at the beginning of the program:

- high reliability                                      (not yet quantified)

- comfort and convenience
  front and rear seat comfort
  noise level                                           (not yet quantified)
  suspension
  efficient heating, ventilation and
       air-conditioning                                 (not yet quantified)
  good controls and displays
  comfortable passenger room
  good useable luggage room

- on-the-road performance
  start easily                                          (not yet quantified)
  run smoothly
  ride smoothly (w. full load too)
  smooth shifting
  quality brake system                                  (not yet quantified)
  comfortable turning radius

- ease of maintenance
  improvements in current state-of-the-art built-in-testing and fault isolation

- safety
  crash hardness: driver, passengers       (not yet quantified)
  dependable air-bag system
  antilock brakes
  comfortable safety belts
  compatible with child safety seats
  quality tires

- Electric battery
  life       5 yrs
  cycle life       600
  cost       $100 (US$)/kilo-watt-hour
  recharge time       8 minutes for 50% re-charge
  recharge frequency (miles)       80 miles
  ability to support power demands of accessories

- Warranty       5 years (conditions yet to be determined)

- Most desired-accessories

### Management Approach

The program will be managed by a new venture company in accordance with partnership agreements documented separately. The T & E personnel assigned to the program will operate as an integral part of the design and development team. The T & E manager will report both to the program manager as well as directly to the company Board of Directors. He will be given control of all T & E resources needed to execute the T & E as described in the most recently approved version of this T & E Master Plan.

At the end of each phase of design and development, the program manager, T & E manager, the Quality Assurance manager, the design manager and the production manager will review progress to date against the guidance and criteria of this plan to confirm adequate completion of the prior phase and readiness to enter the following phase. Concurrence by all constitutes a Decision-To-Proceed (DTP). Four DTP checkpoints for such reviews are planned, as shown in Figure D.1. If any one of the five does not concur with the adequacy of progress, the company president will appoint an independent team to assess the progress and recommend a resolution, or will make that determination himself. T & E exit and entry criteria will be separately identified for each phase of the program development.

## C2.2 – Integrated Schedule

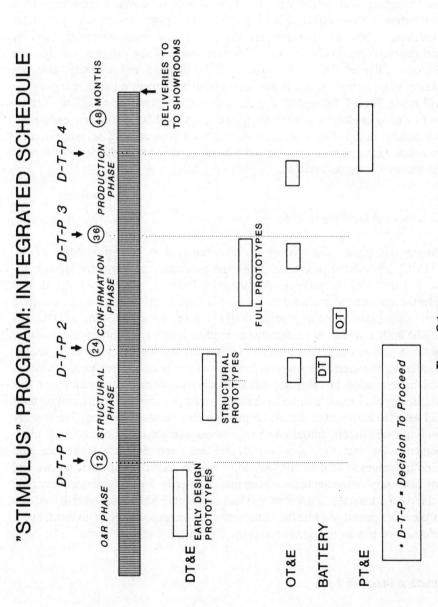

**Figure C.1**
– Integrated Schedule

# C2.3  – Developmental T & E

*Overview*

The program will begin with the design and structural engineering of a conventional automobile, reflecting what marketing surveys and past sales experience show are the features the public are most interested in. Few performance criteria are known at this early stage of the program. As they are firmed up, the related T & E of each will be planned and integrated into this Master Plan. In the meantime, research and development on the battery system will be conducted. Estimates of space and interface requirements of the system will be used in early structural design work and will be continually updated as the battery design firms up. Surrogate power sources will be used in early structural prototypes of the auto until more realistic preproduction versions of the batteries become available.

*Objectives and Requirements Phase*

During this phase, the overall performance and style requirements of the STIMULUS will be determined with the assistance of computer aided engineering and market surveys. Performance features will be prioritized, and selected ones will be included in an update of this Plan – along with quantitative goals and objectives as appropriate. Early Design Prototypes of the STIMULUS will be built and used to confirm the economic design and production feasibility of these performance requirements. Different combinations of vehicle length, width, height, wheelbase, cargo volume, front and rear headroom and leg rooms will be evaluated in computer based models for comfortable seating of five adults. Physical models of up to the three most promising designs will be built and tested to confirm the outputs of the computer aided engineering. Preliminary cooling assessments, initial crash worthiness assessments, powertrain calibration/emission and fuel economy evaluations, and evaluations of ride and handling, climate control and durability will be made. When that testing and any necessary refinements have been made, a Ready-To-Proceed review will be held. Also during the Objectives and Requirements Phase, detailed risk analyses of the more promising electric battery system designs will be conducted, and a selection of one for integration into the STIMULUS will be made.

*Structural Design Phase*

Detailed design of the automobile body will be accomplished, recognizing that the interfaces with the new electric battery system will not be well known and must be kept as flexible as possible. Twelve structural design prototypes will be built to the specifications that result from the Early Design Phase. They will be built for T & E to confirm design features, for reliability development, for early accelerated life testing, for further crash hardness tests, and for breaking system tests. The Structural Prototypes will not have the battery system and the power train. The power train operation will be simulated, if necessary, for the auto structural testing. T & E will be conducted on the breaking systems, air-conditioning system, defrosting and defogging systems, engine cold and hot starting, tire aging, and vehicle paint and rust protection coatings. Six prototypes will be built for crash tests.

### Confirmation Phase

The electric battery system will be integrated into the design, the remainder of the design will be completed (e.g. body styling, interior, options), and full-up prototypes of the STIMULUS will be built for T & E to verify that the completed design meets specifications and program objectives, and is ready for mass production and sales.

## C2.4  – Operational T & E

### Overview

Independent Operational T & E is that T & E which will be conducted (1) to verify that the vehicle has met selected T & E design requirements and commonly accepted T & E standards and (2) to get early customer feedback.

### Confirmation of Design Requirements

At the Nevada Automotive Test Center, vehicle dynamics T & E will be conducted on a wide variety of test courses and terrains at the Nevada Automotive Test Center. To the extent that capabilities for environmental tests (high- and low-temperature operation, salt, fog, relative humidity, solar radiation, rain, blowing sand, and dust) do not exist or are not available at a reasonable cost at the facilities of the joint-venture company members, the NATC will be used for these tests also.

*Full-up Prototype T & E*

When full-up prototypes are available, they will be used to complete or repeat tests at the Nevada Automotive Test Center, and for customer feedback evaluations. A cross-section of customers reflecting typical users will be given prototypes to operate and maintain for varying periods of time. Some evaluations will be timed to end before actual sales, allowing an opportunity to make fixes to the early sales models. Some of those fixes may be back-fit into the prototypes being operated by those whose evaluations are longer term, and will be scheduled to end after sales have already begun. Two full-up prototypes will be provided to the Vehicle Research Center for a range of head-on and offset crash tests.

*Early Customer Feedback*

As part of the five-year warranty of vehicles sold during the first year, customers will be required as a condition of continuing the warranty to submit the vehicle for a thorough one-year check-up, and to participate in written customer feedback evaluations developed by the manufacturer. In addition, in selected high-volume dealerships, customers will be given incentives to participate in carefully structured group discussions about the vehicles, conducted by design and marketing people from the joint-venture company.

## C2.5   − Special Resources

The following resources are those which will require special attention to obtain. At this time, none are quantified:

(1) Finances, including government grants (yet to be determined)

(2) Number of prototypes:

| | |
|---|---|
| early design | 6 |
| structural | 12 |
| full-up | (yet to be determined) |

(3) Services of Nevada Automotive Test Center

(4) Services of Vehicle Research Center

(5) Services of facilities at the four parent companies.

(6) Cost of customer feedback program.

— END OF T & E MASTER PLAN —

# C3 SUMMARY

The following points emphasized earlier about a T & E Master Plan are illustrated in this notional Plan for the STIMULUS:

- First, the TEMP exists, even though it is so early in the program that few quantitative performance objectives have been set. That fact reflects an obvious commitment on the parts of the managers of the program – all of whom are signatories to the TEMP – to make the T & E program visible and a visible tool for engineering design, for programmatic decisions, and for customer feedback.

- Second, the way the key requirements are shown, it is clear that the T & E program is and promises to remain continuously flexible, open to readily being tailored to what the engineering studies and financial resources reveal is in the best interests of the program. The analysis of major risks in paragraph C2.1 is a key indicator of where management has set the program priorities, allowing those planning and executing the lower-level T & E planning to understand what top management is thinking, and to make their judgments accordingly.

- Third, the overall development program is structured in such a way that it is being paced by T & E results. Four major checkpoints, called 'Decisions-To-Proceed', are included in the program schedule at which meaningful T & E results will be available and be used in the decision to enter the next phase of design.

- Fourth, the T & E program shows good teaming throughout. The major departments are all signatories to the TEMP. Most would be involved in each 'Decision-To-Proceed'. And the teaming includes bringing the customer in early.

- Fifth, operational T & E to be conducted to confirm achievements of requirements and to provide customer feedback is included. It is given much visibility by treatment in a separate part of the TEMP. This would send a message to everyone that there is a strong commitment to conduct such T & E in this program and to use the results in key decision making.

- Sixth, the TEMP is written at the top level. Although there is not much detail regarding quantitative performance requirements in this early version of the Plan, it appears that the requirements to be tracked at this level of the program will indeed be the dominant ones, the ones which, if they are not met, will jeopardize the success of the program.

When used properly, the TEMP has been found to be an almost indispensable tool for planning the overall T & E in development programs for complex systems and products.

# Appendix D

## The International Test and Evaluation Association

The International Test and Evaluation Association (ITEA) was founded in 1980 as a non-profit professional organization dedicated to furthering the professional and technical interests of the T & E community. Marked by consistently steady growth, ITEA is now the leading association for professionals in the field of T & E.

ITEA engages in a variety of educational activities. It sponsors symposia, seminars and workshops; publishes a quarterly *Journal of Test and Evaluation*, and supports scholarships at several universities. In addition, chapters worldwide provide members with diverse programs designed to expand their knowledge of the field. Since the late 1980s, ITEA has maintained an annual awards program recognizing individuals who have made significant contributions to T & E.

ITEA headquarters is located at:

4400 Fair Lakes Court
Fairfax, Virginia 22033-3899
USA

Its telephone number is 01-703-631-6220.
The following is a list of the local ITEA chapters and their Presidents, many of which chapters conduct their own programs of monthly meetings and seminars:


**AUSTRALIA**

Southern Cross Chapter
Mr Viv Crouch
Australian Centre for Test & Evaluation
University of South Australia
Salisbury Campus
Salisbury East
SA 5109

**CONNECTICUT AND RHODE ISLAND**

Narrangansett Bay Chapter
Mr Robert Ricci
Naval Undersea Warfare Center Code 3892
Newport, Rhode Island 02841

**MASSACHUSETTS**

New England Chapter
Dr Albert Graff
Raytheon Services Company
2 Wayside Road
Burlington MA 01803

**NEW JERSEY**

New Jersey Coast Chapter
Mr Seymour Krevsky
69 Judith Road
Little Silver
New Jersey 07739

South Jersey Chapter
Paul J. Wolownik
FAA Technical Center ATQ-3
Atlantic City International Airport
New Jersey 08405

**DISTRICT OF COLUMBIA &
VIRGINIA**

George Washington Chapter
COL Kenneth Spencer USAF
Joint Program Office for T & E
1420 Menoher Drive
Andrews Air Force Base
Maryland 20331

**MARYLAND**

Francis Scott Key Chapter
David Robinson
Dynamic Systems Inc.
PO Box N
Aberdeen MD 21001

Southern Maryland Chapter
Mr Sandy Woodard
DynCorp
189 Shangri La Drive
Lexington Park MD 20653

**OHIO**

Wright Chapter
Mr Dennis E. Crouch
PO Box 31663
Dayton OH 45437

**VIRGINIA**

Tidewater Chapter
Mr Michael Field
Eagle Systems, Inc.
468 Viking Drive
Virginia Beach VA 23452

**ALABAMA**

Rocket City Chapter
Mr Jack Bissinger
917 Forrest Heights
Huntsville AL 35802

**FLORIDA**

Central Florida Chapter
Mr Henry I. Jehan, Jr
113 East Greentree Lane
Lake Mary, FL 32746

Emerald Coast Chapter
Mr Dick Bacca
Sverdrup Technology, Inc.
PO Box 1935
Eglin AFB, FL 32542

## GEORGIA

Atlanta Chapter
Mr Raymond Jones
Rockwell International
Tactical Systems Division
1800 Satellite Blvd
Duluth GA 30136

## TENNESSEE

Volunteer Chapter
Mr Tom Best
AEDC/XRR
Arnold Air Force Base
Tennessee 37389

## COLORADO

Rocky Mountain Chapter
Mr Larry Gray
National Systems Research Company
5475 Mark Dabling Blvd
Suite 200
Colorado Springs CO 80918

## ARIZONA

Ft Huachuca Chapter
Mr William D. Farmer
BDM Engineering Services Co.
PO Box 2290
Sierra Vista AZ 85636

## NEW MEXICO

Roadrunner Chapter
Mr Russ Foos
SRS Technologies
2305 Rebard Place S.E. Suite 204
Albuquerque NM 87106

White Sands Chapter
Mr James Noble
White Sands Missile Range
STEWS-NRO-CA
White Sands Missile Range
New Mexico 88002

## TEXAS

Lone Star Chapter
Major Harry Pillot, USA
2004 Shadow Ridge
Harker Heights TX 76543

## CALIFORNIA

Antelope Valley Chapter
c/o ITEA
Edwards Air Force Base
California 93523-0296

Central Coast Chapter
Mr Daniel Wenker
30 SW/XP
747 Nebraska Ave, Suite 34
Vandenberg Air Force Base
California 93437-6294

Channel Islands Chapter
James K. Jones
Naval Surface Warfare Center
Code 4L20
Port Hueneme CA 93043

China Lake Chapter
Karl Volland
CTA, Inc.
900 Heritage Drive
Ridgecrest CA 93555

Greater San Diego Chapter
Mr Howard Harvey
SAIC
10240 Sorrento Valley Road Suite 100
San Diego CA 92121

Southern California Saddleback Chapter
Mr Charles Stefan
Interstate Electronics Corp
PO Box 3117 D/4750
Anaheim CA 92803

**HAWAII**

**WASHINGTON**

Mid-Pacific Chapter
Mr Paul Molyneaux
ITT Federal Services Corp
PO Box 1225
Kekaha HI 96752

Pacific Northwest Chapter
Mr Norm Anderson
PO Box 3999
Mail Stop 8H-11
Seattle WA 98124

# Further Reading

Dvorak, Mark and Equid, Rod, *Principles of Test and Evaluation*, course notes, Australian Centre for Test and Evaluation, University of South Australia, November 27, 1995.

Henley, Ernest J. and Kumamoto, Hiromitsu, *Reliability Engineering and Risk Management*. Prentice-Hall, Englewood Cliffs, N.J., 1981.

Jones, James V., *Engineering Design: Reliability, Maintainability and Testability*. TAB Books, Blue Ridge Summit, PA, 1988.

Locks, Mitchell O., *Reliability, Maintainability, and Availability*. Hayden Book Company, Rochell Park, N.J., 1973.

Meister, D., *Behavioral Analysis and Measurement Methods*. Wiley, New York, 1985.

Smith, John M., *System Engineering Management Tools*. University of South Australia, Australian Center for Test and Evaluation, August 1995.

# Index